JACOBS BEACH

Kevin Mitchell is the boxing and tennis correspondent for the *Observer* and the *Guardian*. He is the author of *War, Baby*, which was shortlisted for the William Hill Sports Book of the Year, and the co-author of Frank Bruno's autobiography *Frank*, which won Best Autobiography at the British Sports Book Awards.

Also by Kevin Mitchell

War, Baby: The Glamour of Violence
Frank: Fighting Back (*co-writer with Frank Bruno*)

JACOBS BEACH

The Mob, the Garden and the Golden Age of Boxing

Kevin Mitchell

Yellow Jersey Press
LONDON

Published by Yellow Jersey Press 2011

2 4 6 8 10 9 7 5 3 1

First published in Great Britain in 2009 by
Yellow Jersey Press
Random House, 20 Vauxhall Bridge Road,
London SW1V 2SA

www.randomhouse.co.uk

Addresses for companies within The Random House Group Limited can be found at: www.randomhouse.co.uk/offices.htm

The Random House Group Limited Reg. No. 954009

A CIP catalogue record for this book is available from the British Library

ISBN 9780224075091

The Random House Group Limited supports the Forest Stewardship Council® (FSC®), the leading international forest certification organisation. All our titles that are printed on Greenpeace approved FSC® certified paper carry the FSC® logo. Our paper procurement policy can be found at: www.randomhouse.co.uk/environment

Printed and bound in Great Britain by CPI Bookmarque, Croydon

For Marie

CONTENTS

ILLUSTRATIONS

Section One

Section Two

1 Rocky Marciano lands the knockout blow that finished Joe Louis's reign, in October 1951 (courtesy of Popperphoto/Getty Images).

2 Senator Estes Kefauver from Tennessee in 1956 (courtesy of Bettman/ Corbis); the hands of crime boss Frank Costello during the Kefauver Senate hearings into organised crime(courtesy of Time & Life Pictures/Getty Images).

3 *Crime in America* by Estes Kefauver, published in 1951; mobster Benjamin 'Bugsy' Siegel shot at his home in Beverly Hills in June 1947 (courtesy of Bettman/Corbis).

4 Mobster Frank Costello poses at the Beverly Club in New Orleans in March 1949 (courtesy of Time & Life Pictures/Getty Images); Blinky Palermo puts the world welterweight title belt around his fighter, Johnny Saxton, after beating Kid Gavilan in October 1954 (courtesy of Bettman/Corbis).

5 James D. Norris at the fights in 1951; Truman K. Gibson Jr, interviewed by the US Supreme Court in January 1959 (both courtesy of Time & Life Pictures/Getty Images).

6 Reporters gather around the hospital bed of Los Angeles promoter Jack Leonard in 1959; Frankie Carbo leaving court in 1959 (both courtesy of Bettman/Corbis).

7 Don King on the telephone, Miami Beach, Florida, in 1977; world heavyweight champion Charles 'Sonny' Liston with his manager, Pep Barone, in 1960 (both courtesy of Bettman/Corbis).

8 Liston is beaten by heavyweight champion Muhammad Ali in round one in Lewiston, Maine, in May 1965 (courtesy of Bettman/Corbis).

1

THE BEAST WITHIN

The story you are about to read has a beginning and a middle, but no end. It is a story about the fight game, and the fight game is an unkillable beast. What it did yesterday, it does today and, unless the sun doesn't rise somewhere, it will do the same tomorrow.

Some periods and places, though, live in the imagination more vividly than others. The fifties were such a time, New York such a place. While no age exists in isolation, there is a backstory to the fifties that makes those years unique in boxing. In that decade in that city, in a venue that has been the spiritual home of the business for more than a century, a coterie of chancers came close to doing the impossible: they nearly killed the fight game.

A lot of people were responsible for what happened in and around Madison Square Garden in those ten years: gangsters, promoters, managers, TV moguls – and some of the fighters.

Jake La Motta, for instance. Jake was the less-than-beautiful bull born on the Lower East Side of Manhattan, a crude, tough kid who raged through his era with manic energy, hounded by the Mob and, very occasionally, his conscience. He was a wife-beater, a rapist and thief, a mugger and liar who went on to become a raconteur skilled in reheating his past. He was then and remains today an extraordinary man, a fighter who struggled to ever say sorry, who expected no apologies in return and, crucially for the making of his legend, clung to the notion that he wouldn't go down. The words he is famously supposed to have uttered through purpled lips in Chicago on St Valentine's Day, 1951, while enduring a barely legal beating at the expert hands of Sugar Ray Robinson, were, according to Martin Scorsese's evocative take on his life, *Raging Bull*, 'You never got me down, Ray.' Those words stand as a boxer's battlecry of futile pride, even though, as his biographer Dave Anderson revealed years later, Jake never said them.

Nevertheless, Jake La Motta is the nearest thing to an animal boxing has ever seen. Stumpy, strong, bobbing up and down as if moving through the jungle in search of food, Jake stalked his prey, splayed-legged, as his short, hairy arms carved a wide, venomous arc, and with only cursory regard for his own health. It was fighting without artifice. It lacked any self-consciousness and it was driven by a mix of bravery and foolishness. But to disregard the defensive tenets of the sport was to court disaster. No man could go in six times with Sugar Ray Robinson and think he would emerge unscathed. No man except Jake La Motta. But, as he was to learn in five of those encounters, he was as human as the rest of us. He was not a beast, after all. And, of course, La Motta did go down. Not that night, but on others against much lesser fighters. More than once. They all did. One way or another, literally or metaphorically, everybody takes a count.

It is central to boxing's myth that the fighter is king, that the great ones rule through the power of their fists, the strength of their chins and the fortitude housed in abnormally large hearts. But they are all blood and bone. La Motta was the essence of the powerless fighting man. He was proud but corruptible, not only because he had to bend to the will of the people who controlled his life, but because, as hard as he was, even he had to admit to his physical and spiritual vulnerability in the end. Sugar Ray couldn't put him down; but boxing did.

La Motta epitomises boxing's gift for crushing truth to a pulp. Although Jake was a hell of a fighter, one of the best middleweights of all time, he had no problem in sustaining his legend by denying for more than a decade that he had thrown his fight against the mobbed-up Billy Fox at the Garden in 1947. If he had confessed when he was first accused of the fix, he would have been banished and forgotten. His silence earned him his shot at the world title ten fights later, in 1949 against the exquisite Frenchman Marcel Cerdan. It also lent La Motta a twisted immortality. It wasn't until a Senate hearing in 1960, however, six years after he'd quit boxing, that La Motta came clean. By then, he was on his way to becoming a parody of himself, albeit a famous and rich one, to be remembered eventually as the Raging Bull.

La Motta is a small part of this tale, and his experiences are not unique. But Jake (and his literary collaborators) had a better handle on it than most. As he was once quoted as saying, 'I also noticed that around the gym all the time there were the Mob guys, for the very simple reason that there's always betting on fights, and betting means money, and wherever there's money there's the Mob. If you paste that inside your hat it will explain a lot of things to you and maybe even save you some trouble.'

He would know.

La Motta's time and place stretched from pre-war New York and other fight towns on through the fifties, a decade in which the Mafia – historically voracious movers of meat in the fight game – wanted to control him and every other worthwhile pug of the era. They carried out most of their shenanigans in the Garden, a place that housed as much deception as it did heroism. That particular establishment is the most famous of the four to have worn the name on various sites in Manhattan, never more than a cheroot spit from Broadway, since 1879. For seventeen of his 106 fights, it is where La Motta established his reputation, on good nights and nights when he didn't look as good as he might have done.

Other men, characters who never threw anything more threatening than a glance, were the real rulers of the ring, not the likes of La Motta. Professional boxing would not survive without either the compliant fighter or the scheming moneymen, and this story is full of them: the old Garden promoter Mike Jacobs, Mafia servants Frank Costello, Frankie Carbo and Blinky Palermo, along with their more presentable business associates, Jim Norris and Truman Gibson, plus a large cast of faceless cronies. These were the men who ran boxing for a generation, the black-and-white days either side of the last world war, mainly in the atmospheric old auditorium situated on Eighth Avenue between 49th and 50th Streets, in the middle of the most exciting city in the world.

And at the black heart of boxing's empire was Jacobs Beach. It's gone now but for twenty years it was the only place to be for boxing's tsars and their camp followers. Geographically, it covered the stretch of pavement on West 49th Street between Broadway and Eighth Avenue, with the Garden and Jack Dempsey's Restaurant at either end of the block. Around about the middle was 225 West 49th, the ticket office of Mike Jacobs, the one-time scalper after whom it was named. For nearly two decades he decided who did and who didn't fight in the Garden. From his centre of the boxing universe, journalists, gamblers, managers, underworld goons and hangers-on would walk across the street to the Forrest Hotel – the occasional residence of Damon Runyon, not to mention women of dubious moral distinction and other denizens of the dark – to talk their dirty business. (The owners sued Bob Hope once for joking that the Forrest's maids changed the rats every day.)

The Forrest – which is now the Time Hotel and a place where out-of-towners some times stay, ignorant of its past, probably – was the place to which most deal-makers had been gravitating since the twenties, guys such as Runyon and his pals, Tex Rickard and Doc Kearns. The charismatic Runyon, who lived for a long time in the hotel's penthouse, was the catalyst for many great gatherings and stories. He loved all sports, but boxing dominated the hushed conversations of the Forrest. The writer Westbrook Pegler wrote in 1936: 'There was always some hungry heavyweight sitting in the big fat chair in the corner, squinting down the street at the clock to see if it was time to eat yet. Sometimes it would be an old, gnarly heavyweight with a dried apple ear and a husky voice from getting punched in the neck. Sometimes it would be a young pink one with the dumb, polite expression that young heavyweights have.'

The label Jacobs Beach – no apostrophe mostly, although, like a fight contract, these things are not set in stone – was given life by Runyon, according to the newspaper columnist Jimmy Cannon. But the American author David Margolick, who has written excellently about the period and about boxing, is not sure. 'More likely,' he says, 'the name came from one of two prime sportswriters of the era, Frank Graham of the *New York Sun* or Sid Mercer of the *New York American*.'

Whatever its genesis, Jacobs Beach lives on only in grainy, nostalgic images. The cartoonist Ham Fisher, who created the long-loved and improbable boxing ingénu Joe Palooka, would include in his syndicated strips the requisite number of zoot-suited sharpies, chewing on old stogies and shooting the breeze with broken-nosed bums hanging about for a break or a broad. That's how Runyon would love to have 'The Beach' remembered, too.

The Beach lasted from 1935, when Jacobs set up his ticket office, until the imagination let it go, at some point in the mid-fifties, a few years after Uncle Mike had been eased from power at the Garden by illness and the International Boxing Club, which was run by his sometime Mob pals. Nostalgia is fine, but sentiment lingers about as long as cigar smoke in boxing. Today, it is just another bit of grubby Gotham pavement. No plaque. No statue.

In 2005, Margolick took Jacobs's niece, Roz Rosee, to the area – and she couldn't be sure where the old rascal's office had been, even though she had once worked nearby in the Brill Building. In fact, it stood on a site now occupied by a garage on the north side of the block. 'He was very smart,' Rosee, then eighty-nine, said of her Uncle Mike, 'but he was not what you would call a gentleman.' You would hope not.

The gangsters who mingled on Jacobs Beach did not go unchallenged in their dealings. What happened there and across the wider landscape of the sport in the fifties was described and dissected by two United States Senate inquiries, at the start and finish of the decade. These were instigated by an old-fashioned Democrat lawyer from Chattanooga called Estes Kefauver. When Americans turned on the few TV sets they had in 1950, they saw the upright, earnest senator, a decent man who yearned to be president, trying through the rigour of the law to rid the country of the mysterious men who organised crime, as well as the fight racket. The first of Kefauver's televised investigations – as gauche as the good senator himself – met with qualified success. In the process, however, he became America's original and most unlikely reality-TV star. Every night America saw, for the first time, flesh-and-blood images of the bad men Kefauver was trying to put away.

The wise guys who owned boxing also owned the heart and soul and much of the bank account of the heavyweight regarded then as the best there ever was. Joe Louis, the champ everyone loved, was a balding shadow with debts, a failing marriage and a mind in decline by the time he sold the title to the IBC. Kefauver was as knowledgeable a boxing fan as most American males of his time and was saddened by the spectacle of Louis passing on the championship he had once valued so highly. Despite the best efforts of the senator and other do-gooders, though, the Mafia continued to control most of boxing's significant world titles for several years. And, sad as it is to acknowledge, it was Joe Louis, his faculties already shredded, whose gentle nature and lack of financial alternatives made it possible.

Not every fight in the Garden in that decade was a predetermined result. But regular punters went there whispering. Often they trudged home scowling, stepping over ripped-up betting slips as they made their way towards the subway station on 50th Street, for the old Eighth Avenue Independent or the West Side IRT. Many a journey home from the Garden would be filled with loud discussions about the antecedents and social connections of the judges and referees. And nearly every discussion of a night at the fights was peppered with two words: the Mob.

The better heeled among New York's fancy stayed on, for champagne, music and whatever version of risqué entertainment they could safely be seen to be patronising. Of the swanky nightclubs favoured by the silk-scarf set who slummed it at the Garden were El Morocco on nearby 54th Street and the more formal Cafe Lounge in the Savoy-Plaza Hotel, at Fifth and 59th. There was the mildly notorious Kit Kat Club, on 55th, 'Harlem moved downtown', as guides of the time described it, and 'popular in the early hours of the morning'. There was the Stork Club on 53rd, expensive, packed with New York's aristocracy and the perfect place to mull over wages won and lost at the Garden. There was also, of course, the Cotton Club, on Broadway and 48th, long-time haunt of boxing's inner sanctum; Mob connections with the Cotton stretched back to the twenties, and the neon signs that flashed outside advertising '50 Tall, Tan, Terrific Gals' said it all.

And, for the hardcore, there was always Toots Shor's, the ultimate post-fight den, parked at 51 West 51st Street, a short stroll from the Garden. It was here that many great stories were born (some of them true) and just as many reputations ruined – or enhanced, depending on your view of life. The actor Jackie Gleason, large in every way, threw his weight around here a lot, brawling and boozing until he collapsed on the floor. On more than one occasion the proprietor stepped over him with practised nonchalance. That was Bernard 'Toots' Shor, a rabble-rousing adventurer from Philadelphia, who was big and ugly enough to earn a living as a speakeasy bouncer in the fading years of Prohibition. He met and liked Damon Runyon, whose clout gained Toots entry to New York's demi-monde. Toots had found his natural home and, in 1940, opened his eponymous establishment. Like Jacobs Beach, it would live on past a point of dignified closure, a tatty relic in the end.

In the age before rock and roll, before TV took a grip, before the Pill, as the horns swung from big band to boogie to blues, New Yorkers loved a simpler version of the boxing universe: they loved Toots, the Garden, great fights, as well as the smell and thrill of Manhattan's throbbing ribbons of light. For a while, there was no place else to be for those in search of sanitised depravity.

It was in these clubs and bars and coffee shops that men in their thin-brimmed pork-pie Adams fedoras, tugging on their Romeo Y Julietas, stinking of Pino Silvestre cologne, talked quietly with men whose flat noses betrayed their calling. These were not innocent men. They were players. They made boxing what it is, for better or worse.

The Mob had been deep in the heart of boxing since the early days of Prohibition. The handing on of power from one set of thugs to the next had not been seamless, but it had been unstoppable. By the late forties, the International Boxing Club, run at one remove by Carbo, had slipped without ceremony on to the throne. Usually one step ahead of the law, they controlled boxing late into the fifties, by which time complacency would weaken their grip on the business and on reality. When Kefauver went after the Mob one last time in 1960, he came closer to delivering a knockout blow than he and other agencies had in previous efforts. This was partly through the belated confession of La Motta, but mainly was down to the staggering hubris of the bloated and arrogant warlords who reckoned they could continue to rule the lives of thousands of people in the fight game for as long as they chose to. Even as they sneered at Kefauver's righteousness, the wise guys were compelled to acknowledge Joe Louis's own ring maxim: you can run, but you can't hide.

Those characters and those joints are memories now. But La Motta lives on (at the time of writing), still telling his stories. His contemporaries have nearly all gone down for the count, replaced by facsimiles. Don King, who started his boxing journey in the fifties, carries on the fine tradition of hucksterism that began with P.T. Barnum in the first Garden and was refined by the likes of Rickard, Kearns and Jacobs in the others.

What the Mob did between 1950 and 1960 at the Garden provides a snapshot of a sport and business many people said, even then, was out on its feet. They've been saying that since the days of Jack Broughton, Jem Mace, John L. Sullivan, Primo Carnera . . . They're saying it now. Somehow, against the odds, boxing keeps getting off the floor.

2

THE RING IS DEAD

There was a party in the Garden in September of 2007, but, for once, Muhammad Ali couldn't make it. He was a palsied shadow of himself, sitting quietly at home in Louisville, sixty-five years old, tended by his wife and nurse, Lonnie, and informed of the story second-hand by the friends who invariably descended upon him whenever boxing hit the news. How he would have loved to make the trip to New York. It was there, thirty-six years earlier, that he and Joe Frazier had given the Garden and the world one of their sport's most enthralling contests, the Fight of the Century. But that was just an entry in his scrapbook now for the man who saved boxing, as the news filtered through about the end of an era.

What appeared in Ali's newspaper in Kentucky that morning was a bulletin from the Associated Press, issued at 4.35 p.m. (Eastern time) the previous day, the 18th. Datelined Canastota, New York, it read: 'After 82 years, Madison Square Garden will retire its storied boxing ring and donate it to the International Boxing Hall of Fame, where it will go on display this fall.'

This was the ring in which La Motta first fought, and lost to, Sugar Ray Robinson, in 1942.

This was the ring in which Rocky Marciano knocked out Joe Louis in the Brown Bomber's last fight, in 1951.

This was the ring where Willie Pep and Sandy Saddler threw the rule book away and brawled in 1951.

This was the ring where Randolph Turpin brought the crowd to their feet in losing heroically to Carl Bobo Olsen in 1953.

This was the ring where Joe Louis was lucky to get the decision against Jersey Joe Walcott in 1947, and where Billy Graham was robbed against Kid Gavilan in 1951 – as was Lennox Lewis against Evander Holyfield in 1999.

This was the ring where Lou Ambers, Tony Canzoneri, Beau Jack, Dick Tiger, Ken Buchanan, Roberto Duran, Fritzie Zivic, Ike Williams, Joe Frazier and Muhammad Ali thrilled, shocked and amazed us.

This was some ring.

Now, in one short sentence devoid of sentiment – as is the detached way of wire services – a collection of nuts and bolts was to be buried, not without ceremony but with little regret outside the fight community.

The following morning in New York, a big man, wearing a smile permanently wreathed in sardonic double meaning, stood in the middle of the pensioned-off ring and boomed: 'The ring is dead! Long live the ring! Heh! Heh!' The pre-eminent salesman of his or anyone else's time, Don King, as ever, had found quotable pith with which to put his stamp on an upcoming promotion in the Garden. There would be one last fight in the old ring, he said, and then we could start spilling blood in a brand-new one!

As it happens, there would not be another fight in the old ring. One of the proposed antagonists, Oleg Maskaev, was injured and the fight was called off. Another boxing mirage.

King loved the Garden, not wholly out of sentiment. He was a student of history, and used it constantly to lend his promotions glamour and legitimacy, capitalising on the allure of boxing's past. Certainly, King might be sad to see the old ring go. But he was still standing. At the time of writing, the King is not dead. And we would not wish it so, even if there is a fat parish of enemies out there who disagree with that take on the subject. What a life he's led. What lives he's marred and, to be fair, enhanced. In the late forties, while La Motta and the Mob were getting acquainted in New York, King threw a few punches as a skinny teenage flyweight back in Cleveland. He won a couple then quit the sport after being knocked out in his fourth bout. Like the guy who stole Cassius Clay's bicycle in Louisville and led him to take up boxing as a bullied adolescent, the long-forgotten schoolboy boxer who put Don King's lights out lives anonymously forever in boxing's hall of myths.

When King gave breathless birth to his valedictory over the sacred ring in Madison Square Garden – one of hundreds for which he will be remembered when he is eventually laid to rest by gout or universal *Schadenfreude* – the roped square was on its way to the International Boxing Hall of Fame in quiet Canastota. Rebuilt and revered, it resides as a reminder of the skulduggery and high times that took place in what for most of the twentieth century was the most revered arena in sport.

Nobody can be sure how many fighters stepped through the ropes of the most famous Garden ring over the course of eighty-two years. We can be certain, though, that the last title fight there was on Saturday 9 June 2007, when Miguel Cotto of Puerto Rico stopped Zab Judah of Brooklyn in the eleventh of twelve scheduled rounds to retain his World Boxing Association welterweight belt in front of 20,658 fans, most of them New York Puerto Rican supporters of the champion. It was the biggest crowd the venue had seen for a championship bout outside the heavyweight division.

The fight was worthy of the surroundings. The ghosts of the Garden would not be disappointed, by either the class of the winner or the bloody-minded courage of the loser. This dirty, glorious space had always celebrated heroics and, when required, drowned perceived tankers and bums in a hail of derision. Now, as worn out and used and useless as a washed-up fighter, the ring was being laid to rest for good.

To fighters, the ring was a place of work; to admirers of architecture and engineering, it was a work of art. It was a minor marvel from a time when detail and artisanship mattered. The brass was polished so assiduously, it is claimed, that, when TV cameras fell on it in the fifties, executives complained there was too much glare for the cameras – much as they had pointed out to the advisers of Dwight D. Eisenhower that his shiny dome was a distraction to both the cameraman and the electorate when he ran for the presidency in 1952.

The ring wasn't always on TV, though, and it wasn't always fixed in place in the Garden. It was moved about like a shrine, from the Garden to Yankee Stadium, the Polo Grounds, even a gym in Little Italy. But its home was on Broadway.

It measured 342 square feet, 18ft 6in on each side inside the ropes – smaller than today's 20ft by 20ft rings – weighed more than a ton and was held together through a complex set of 132 interlocking joints. A lot of heads hit the canvas (some more willingly than others), which was replaced periodically, as were the padding and the ropes, up against which wily veterans would scrape the backs of bright-eyed novices.

It was the fighters' stage, where no man could lie to himself for very long (unless paid to do so). The canvas, ropes and posts are as blessed in boxing as the altar is in religion – a place of worship, and, more often than some people would like to admit, a place of sacrifice.

Why a ring, why square not circular? The ring is the accidental invention of the Georgian bare-knucklers who stepped on to whatever patch of grass was available and far enough from the unwelcome attentions of the law to accommodate the bloodlust of the Fancy. There they'd face off, in deepest Surrey or Hampshire, maybe Kent or Bristol or Yorkshire, surrounded by four wooden posts and some rope, erected not for any legislated purpose of keeping order between the pugilists, but to hold at bay the intoxicated mob. It was square only because the prizefighters' seconds stood opposite each other and those entrusted with policing the occasion would put a stake beside them, upon which they'd place their coats and hats; to run the rope around the fighters, it made sense to have two more supporting stakes, on the other diagonal, and on these were placed the bets, or stakes, in the care of some hopefully reputable third party. Thus the square ring simultaneously became geometrically incongruous and indestructible in the imagination.

And so they gathered in the Garden one last time to pay homage to an inanimate object, with the very animated Don at the centre, the fighters, as ever, all around the man who sometimes made them rich and celebrated, sometimes poor and discarded.

Something more subtle than a King monologue was at work that autumn of 2007. There was a case for tearing down the old ring, certainly; it was starting to creak dangerously. But so was boxing. This was more than the transfer of some metal, wood and canvas from New York to a museum in a small upstate town. The reality was that the sport was in trouble – and now another piece of the fragile edifice holding it together had been stripped and consigned to a museum. Represented as regeneration by interested parties, taking down the ring also symbolised the dismantling of the fight game. It would not end there. Even as the carpenters were packing the ring into crates, architects, engineers and lawyers at nearby City Hall were talking seriously about the demise of an even more obvious boxing institution: Madison Square Garden itself.

The one still standing, the one from which the ring had been plucked, is the fourth Garden. It had been built over Pennsylvania Station, on Eighth Avenue between 32nd and 33rd Street, forty years before and had operated since 1968. Now, it seemed, as part of the endless odyssey, there might be a fifth Garden. On Tuesday 23 October 2007 – five weeks after the ring had been dismantled – the Empire State Development Corporation unveiled a $14 billion plan to level the old building and put up a new one nearby on Ninth Avenue. Will it happen? Nobody knows. As ever in boxing, we will have to wait until fight night.

The ring is dead. Long live the ring.

3

NEVER FAR FROM BROADWAY

There never was a fight promoter more suited to his trade than Mike Jacobs. He started life in gaslit New York City in 1880, one of eleven kids in a family of Jewish immigrants from Dublin, and never took a backward step as long as he lived. Jacobs was born to hustle. His mother and father had stopped off in Ireland when fleeing religious persecution in Eastern Europe, and, after they had joined the Irish rush to the New World, Mike grew up as a cultural oddity in the Hibernian ghettos of the Lower West Side. He was resourceful, unsentimental and hungry. He sold candy on the boats that went to Coney Island and, from the age of twelve, he scalped tickets outside the second Madison Square Garden. Fans looking for admission to a fight at the last minute any time in the 1890s would find the skinny kid with the loud mouth striking the hardest bargain. He was ruthless in his negotiations. Young Mike could turn a $2 ticket into a $10 profit in the twinkling of his Irish-Jewish eye. There wasn't a better Fagin on the streets of the city. 'After sixteen, I was never broke again,' he said once.

Jacobs was so good a salesman that, in a lifetime of aggravation and conflict, he was always confident of a result. Win or lose, his demeanour did not change much. He rose to the top of boxing's dungheap as if by divine edict, and those who looked to outsmart him could not penetrate an exterior born of adversity and forged in greed. Jacobs died a rich if unsmiling man. While it was a love of money rather than the sport that drove him, nobody questioned his right to be there. He was one of the army of foot soldiers who made boxing tick, if not always after the fashion of a tea party.

In an evocative piece written in 1950, Budd Schulberg described him as the 'Machiavelli on Eighth Avenue'. Other sports writers called him 'Monopoly Mike'. Dan Parker, the most perceptive and hard-hitting of fifties fight writers, named him 'Uncle Wolf'. Jimmy Cannon said Jacobs was 'the stingiest man in the world'. Real enemies, of which there were a few, called him far worse than any of this. He didn't give a damn.

Schulberg saw some good in him. 'He staged 61 championship bouts, promoted 3,000 boxing shows, signed 5,000 boxers, grossed over $10 million with Joe Louis alone, staged approximately 70 per cent of all the bouts below the heavyweight division that grossed over $100,000 (totalling $3 million, with a mass attendance of half a million), attracted in a single year, to 34 Garden shows, nearly half a million people, grossed in that same year $5.5 million, and sold tickets over a 15-year period to more than five million people who pushed at least $20 million through Mike's ticket windows.'

In boxing, it's all about the numbers. Mike Jacobs, who had the heart of an accountant, was the number-one Numbers Man, and the Garden was his bank.

The New York that fashioned Jacobs was some way different from the skyscraper island of glamour we know today. The stench of poverty and sickness haunted Manhattan's poorest, as it had done since the birth of the colony. But every New Yorker, rich and poor, was mesmerised by the bright lights of Broadway.

The thread that links all life in Manhattan was known in the early days of Dutch settlement as Heere Straat, or High Street. Before that it was an established Indian trail called Wickquasgeck Road, running along a prominent ridge of the hilly island. The name Broadway, according to most educated guesses, comes from Broad Wagon Way, and that sounds right.

Boxing and Broadway started their love affair at its lower reaches.

Phineas T. Barnum, a man who advertised his wares with all the subtlety and charm of a hooker, knew how to 'get them in'. He was the original American con man of sport, and he set the tone for the chaos that followed him. From the moment P.T. opened the doors to 'Barnum's Monster Classical and Geological Hippodrome' on the site of an abandoned passenger depot of the New York and Harlem Railroad, at 26th and Madison on 27 April 1874, he embraced the philosophy that became not only his mantra but the guiding principle of the fight game: 'There's a sucker born every minute.'

At the roofless Hippodrome – which *Harper's Weekly* described at the time as 'grimy, drafty, combustible' – the sky was the limit as far as harmless nonsense went. There were waltzing elephants, fire-eaters, the usual sideshow freaks and all manner of proto-Roman excesses, such as chariot races. And fights, many of which were on the level. Looking over proceedings was an eighteen-foot gilded copper statue of Diana, the goddess of love, who, despite her bulk, swivelled in a light breeze, almost tempting God to knock her down in retribution for the sins committed beneath her ample charms. New Yorkers loved the gaudy excess of P.T.'s Hippodrome, but, even then, the foundations and external trappings were shifting.

Before it became properly notorious, the old place had a couple of different names on its way to becoming known generally as Madison Square Garden, in 1879, just a year before Mike Jacobs was born. And here it was that John L. Sullivan created part of his legend. On 17 July 1882, he took on the British fighter Joe Collins, who was known in some quarters as Joe 'Tug' Wilson. Collins/Wilson took up Sullivan's challenge to remain standing for four rounds to collect $1,000. Collins, whose ring history was sketchy and who went down twenty-four times in his efforts to avoid a clean knockout, collected on the dare – but Sullivan's aura was not diminished, at least not among the gullible. They couldn't get enough of this illegal pugilism and the blarney Sullivan brought with it.

The great man's second exhibition, a year later, didn't go so well. Police captain Alexander Williams told reporters he was bringing the entertainment to a halt 'just short of murder'. The following year, Sullivan was arrested at the Garden during his bout with Al Greenfield and charged with behaving 'in a cruel and inhuman manner and corrupting public morals'. As with P. T. Barnum's credo, this was a statement begging to be added to boxing's unwritten constitution.

John L. was king. And, for much of the time, he was on the run from the authorities, like many of his fans. When prizefights couldn't be snuck into the Garden under the banner of education, they were held in fields and on barges. Sullivan was acknowledged as the world's bare-knuckle champion by beating a part-time punch-thrower from County Tipperary, Paddy Ryan, in front of an audience that included the James boys, Frank and Jesse, in Gulfport, Mississippi, in February 1882. Back in New York, the Garden's interest in boxing spluttered along intermittently . . . for a while.

That first Garden was knocked down, rebuilt and repackaged, opening on the same site in 1890. And the new darling of the fight fraternity was an Irish Californian, James J. Corbett, whose *nom de guerre*, 'Gentleman Jim', owed more to alliteration and the vivid imagination of his publicists than any pedigree polished while mixing in high society. When 10,000 sadists flocked through the doors of Garden II on 16 February 1892, Corbett obliged them by beating three men in a row, knocking two of them out cold. Eight months later, Jim was champ and John L. was chump, washed up and in the grip of the bottle. Corbett 'near murdered' the old man and was the new heavyweight king. Gentleman Jim – in the spirit of brotherly love unique to fighters – staged an exhibition in the Garden for the retirement pot of his vanquished foe.

By the time the teenage Jacobs was making a name for himself as a resourceful ticket mover, the intersection of Broadway and Seventh Avenue, at Times Square, laid claim to being the centre of the universe. It was gloriously lit, flashing its temptations twenty-four hours a day. And not many, rich or poor, resisted the temptation to make Broadway and the Garden their preferred place of pleasure.

Opening night at Garden II was special. The vice president of the United States, Levi P. Morton, was among the 12,000 guests who gaped at the temple of vulgarity the eminent architect Stanford White had designed for J. P. Morgan, one of the richest men in America.

'From the principal entrance on Madison Avenue,' writes White's biographer Paul R. Baker, 'the first-nighters moved through a long entrance lobby lined with polished yellow Siena marble, merging into the huge and colorful amphitheater. Gold and white terracotta tiles decorated most interior walls. Two tiers of seats rose along the sides, and three tiers of boxes, trimmed in plush maroon, filled the ends of the vast space. Some 10,000 spectators could be seated comfortably in the amphitheater, and there was standing room or, for some events, floor seating for up to 4,000 more. The high roof was spectacularly supported by 28 large columns. At the center of the roof, an enormous skylight could, as if by magic, be rolled aside by machinery. This was done during the opening performance but it occurred so quietly that most spectators were not even aware of the change until they noticed the cool night air. As in ancient arenas, provision was made for flooding the floor for water spectacles. As at the Roman Colosseum, animal stables to be used for the horse shows and circus performances were placed in the basement below. Here was a bit of ancient Rome, transformed, modernized, and brought to the Gilded Age of New York!'

It was no tent.

By the turn of the century, Jacobs's career had gone from street mischief to seriously influential. Although barely out of his teens, he knew most of boxing's big guys, including Tex Rickard, the Texan who'd spent years making and losing fortunes in his gaming houses in the Klondike. They met in Nevada in 1904, where Rickard promoted his first fight, a world title contest between Joe Gans and Battling Nelson.

When Jacobs got home, he met Bat Masterson, who'd left the Wild West behind him and was going to be a bona fide New York character, like his new friend Damon Runyon. This was an historic coming together of larger-than-life boxing folk.

Garden II, meanwhile, was to entertain them all with the most delicious scandal, one that would set the tone of activities there for a century to come.

Stanford White had a house in fashionable Gramercy Park, a wife, family and a reputation kept clean by an obsequious media. In reality, as Runyon and his pals knew, White was also a New York dandy of substance, with an insatiable libido. While Garden II was his baby, the creation he coveted most was one Florence Evelyn Nesbit, known to all as Evelyn. As befits the story that ensued with the predictability of a naughty nineties melodrama, Evelyn arrived in New York at fifteen, penniless and with a body and face that would buy her all the trouble she and her suitors could handle. She 'modelled' – and won the heart of every man in the city, most notably White and a rival, the cruel and unstable businessman Harry Kendall Thaw.

Against the odds, Thaw won. He married Evelyn and spent his waking hours in a jealous fit. With good reason, it turned out. Thaw put money into a cheap musical at the Garden called

Mamzelle Champagne, which opened on a hot June Monday night in 1906. Five rows from the front sat White and Evelyn, not too cleverly clandestine. Thaw arrived late for the show, drunk, a pistol hanging menacingly from his limp fingers. Without ceremony, he went up to White and shot him dead, through the left eye.

Thaw was sent down and went mad. In so many respects, the murder of Stanford White echoed with metaphors for the fight game. Professional boxing could not exist in a moral vacuum and, time and again, the air hovering over it in Madison Square Garden would be filled with the smell of foul play.

The Great War came and went, devastating a generation. Doughboys came home looking for thrills, of which there was no shortage in New York. And there were plenty of fine writers on hand to chronicle the action. The New York boxing scene has always been sustained – some would say invented – by a rich cast list of literary scallywags.

Jacobs, while never one for books, made sure to stay in with the fight writers, especially Runyon, whom he liked for reasons that had little to do with the music of his words. Runyon, as Jacobs was well aware, was every bit as sharp as he was, tutored in the ways of the world by his father and always on the lookout for a good business opportunity. Jacobs and Runyon would become the firmest of friends.

There is a story, relayed by the fine old New Yorker Jimmy Breslin, which explains how Runyon forged his world view. Breslin (eighty now and still kicking) has it on authority, via Masterson, that Alfred Lee Runyan (the family name's correct spelling) once told his son Damon: 'Son, there will come a time when you are out in this world and you will meet a man who says that he can make a jack of hearts spit cider into your ear. Son, even if this man has a brand-new deck of cards wrapped in cellophane, do not bet that man because, if you do, you will have a mighty wet ear.'

This was Runyon's pedigree. All his life, he moved among

men, and occasionally women, of a gambling instinct. He was particularly close to Masterson. Bat – or Bartholomew, as his mother would have preferred he be called – was a card sharp, gunman of the mythic West, a referee of dubious prizefights and, when he ran out of all those high-class options, a journalist. As for Alfred Runyan, he was a whiskey-wet hack and born liar, an adventurer of the first order who loved the sound of his tales as much as the substance. His famous son, Damon, worked words for a living to rather more lucrative effect. He was a storyteller who bothered not a lot with such fiddling details as the spelling of his surname (he stuck with Runyon after a newspaper got it wrong in his early days) and would go on to bestow on his part of the twentieth century a narrative of consistent unreliability. In young Runyon's genes were the seeds of romance and fantasy. And among the many myths he left us was one for which we should all be grateful: Manhattan.

All in all, you'd prefer to believe Runyon's stories than not. Masterson – for whom the cards fell kindly through the dexterity of his mind and fingers – knew both father and son and liked both, but he was in awe of Runyon the younger, who aspired to be remembered as America's twentieth-century reincarnation of Mark Twain. Masterson believed what Runyon said: that life was mainly 6–5 against, that the little guy always had it tough.

They were all addicted to aphorisms.

'There are those who argue that everything breaks even in this old dump of a world of ours. I suppose these ginks who argue that way hold that because the rich man gets ice in the summer and the poor man gets it in the winter things are breaking even for both. Maybe so, but I'll swear I can't see it that way.'

Those were the words stuck on a slip of copy paper in Masterson's typewriter when they found him dead at his desk on the evening of Tuesday 25 October 1921, in the offices of the *New York Morning Telegraph*.

In all likelihood, Masterson was down at the Pioneer Sporting Club earlier that evening to watch Gene Tunney stop Wolf Larsen in seven. If he were not, Runyon would have had him ringside in any account he wrote. Runyon loved Bat Masterson and everything anarchic and wild he stood for. Years afterwards, he would resurrect his friend as Sky Masterson in *Guys and Dolls*.

Runyon and Masterson are from long ago but, without them, and scores of like-minded characters, the landscape inhabited by those who followed them would have been a rather dull place.

In the fight game, the game of six degrees of separation is a fruitful exercise. Sometimes you don't get to six. It goes like this: Masterson was on hand at several of the major fights of his time, having an intimate association with fighters, managers and promoters whose legacy ran through the business for much of the twentieth century. The man who'd stood side by side with Wyatt Earp in Dodge City and Tombstone (although he missed the infamous Gunfight at the OK Corral) went on to cement a reputation as one of the Wild West's legendary enforcers.

Runyon, besides sitting at the shoulder of boxing's premier entrepreneurs, such as Rickard and Kearns, not to mention all the great and mediocre fighters of his day, helped create the cartel at Madison Square Garden (with the help of his stinkingly rich publishing boss William Randolph Hearst) that would make the Mob's grip on boxing health-threateningly strong. And Jacobs, the urchin from the West Side, was right in the thick of it with all of them.

Into this rich mix came men of suitably dubious character. You can imagine they were not turned away. This was a milieu that relied on a certain degree of laxity in morals. And then, the party-crashers got a helpful little nudge they could hardly believe.

How the Mob got into an unchallengeable position of power in boxing from the twenties until at least 1960 can be laid at the door of two well-meaning fools from America's Midwest. Andrew John

Volstead was a Republican lawyer from the hamlet of Granite Falls, Minnesota, and Wayne Wheeler, of little Brookfield, Ohio, was a stiff-necked, teetotal hick who also went into law and was behind the hugely influential Anti-Saloon League.

In 1920, Volstead, advised by Wheeler, had voted in the 18th Amendment to the Constitution, banning the manufacture and consumption of any drink containing more than 0.5 per cent alcohol. In New York, authorities closed down 15,000 licensed premises. Before they had poured the booze in the Hudson, 30,000 speakeasies had opened their hidden doors. Within five years, that number had grown to at least 100,000.

And still fans thronged to the Garden, juiced up illegally and not giving a damn.

Crime, meanwhile, outpaced the zeal of the crime-busters chasing down illegal booze. And there to cash in were gangsters who now had a sympathetic constituency of millions – ordinary, thirsty citizens who came to view the police with growing irreverence.

'The national prohibition of alcohol – the "noble experiment" – was undertaken to reduce crime and corruption, solve social problems, reduce the tax burden created by prisons and poorhouses, and improve health and hygiene in America,' wrote the American economist Mark Thornton.

Instead, it spawned the most complete expansion of organised gangsterism the world has ever seen. Prohibition gave birth to the Mob as we know it. It changed the moral landscape forever. Legal jobs disappeared. Decent people were driven to crime. What was considered wrong once became the norm. Stealing, casual violence and deceit spread. And, most tellingly, so lucrative was bootlegging, the preserve of the established mobsters, that they turned themselves into businesses. This was the genesis of organised crime in America. The phenomenon grew with names attached: the Syndicate, the Outfit and, chillingly, their dedicated killing unit, Murder Inc.

Variously, the virus-strength spread of brilliantly marshalled illegality has been seen as the work of the Mafia, as well as the myriad ethnic gangs in ghettos all over the country. Really, they should share the credit with Volstead, Wheeler and the bone-headed politicians who voted for Prohibition. But for their puritanical idiocy, we might never have heard of such successful criminals as Al Capone, Bugs Moran, the O'Bannions, all stars of the St Valentine's Day Massacre in 1929.

It's hard to comprehend the impact of this legislation from a distance – except by the startling statistics: murders and serious assault went up by 13 per cent; lesser crimes increased by 9 per cent; prisons bulged by an extraordinary 561 per cent increase in customers.

So, sloshed and wild, New York, indeed all of America, danced the Charleston. They hung from the wings of biplanes that flew over Manhattan. They believed for a while in their own immortality. It was a decade made for Gatsby and excess, all the time pregnant with the certainty of retribution and collapse.

The working classes made heroes of the bootleggers. The Mob seized their opportunity and established such underground hegemony they were virtually untouchable. Americans did not believe, for a variety of reasons (fear, complacency, convenience), that there was anything in it for them to snitch on the criminals who sold them their rum and beer.

A far greater ill visited upon society than sly grogging was the spread of the protection rackets, which instilled fear, even in men of physical courage. Some of those boxed for a living, but their fists were useless against the hoodlums who raked the streets of New York and other cities with sub-machine guns from the safety of their passing Model T Fords.

The lotus-eaters were being driven underground, into the speakeasies, dealing in the dark. Then, at the very time the reactionaries were winning socially, boxing, of all sports, decided to reach for respectability.

Fist fighting in all its forms had, since Georgian days, struggled to stay a step ahead of the law. The National Sporting Club, formed in 1881, regarded itself as boxing's gatekeeper, regulating titles and weights. But boxing grew with such speed after World War I that no private members' club in London was going to contain the ambitions of the trade's rising entrepreneurs in New York.

On the face of it, the urge to cleanse seemed to be spreading from the bars to the ring. The New York State Athletic Commission was formed in 1920 to oversee the Walker Law, a piece of legislation that entertained professional fighting as long as it subscribed to the law's jurisdiction of the commission. In time, the NYSAC established influence over similar organisations in other states – and the world.

Inevitably, however, there were splits from the very start. In 1921, the rest of fighting America set up the National Boxing Association. Anarchy was up and running.

This served only to encourage the mobsters to move in on boxing with saliva dripping through their grins. They did not like regulation, but they did not mind the appearance of regulation – nor its confusing and chaotic replication. This was turf they could exploit and the vultures were quick to land. Arnold Rothstein, the man rumoured to have fixed the 1919 World Series with the help of the former world featherweight champion Abe Attell, would go to the fights then hold court at Lindy's, at Seventh Avenue at 53rd Street, a place where you could get bagels, booze and the skinny on the next big fight. He would sit ringside at the Garden, handing out threats and favours to whoever he chose. In Chicago, Al Capone, a fight fan but bigger enthusiast of making money, bullied his way into the affections of promoters and managers.

Whatever arms were twisted for whatever result, there were still great fights at the Garden – such as the contest in 1922 between Harry Greb, who trained on sex and illegal liquor, and the upright Catholic intellectual Gene Tunney. Gene, who liked to think of himself as a man of letters and who numbered George Bernard Shaw among his friends, was handed his only defeat by Harry Greb, a man for whom reading and writing were not so much chores as irrelevant, except when filling out betting slips. It was a victory for the bad guys. There would be many more.

Young hoods rubbed shoulders with Babe Ruth, who'd moved to New York from Boston in 1920, in time to hit a home run at the opening of Yankee Stadium in 1923 (within ten days of the opening of Wembley Stadium and the White Horse FA Cup Final). These were thrilling, dangerous days, full of adventure for anyone game enough to try his luck with the city's lowlife.

As Yankee Stadium was going up, Madison Square Garden II, White's gauche monument to a bygone age, was about to be reduced to rubble. America was moving on at a furious lick. Nothing was expected to last, except the myths. America lived for today, furiously. The New York sage Westbrook Pegler called it 'the era of wonderful nonsense', and much of it would be played out on the canvas stretched across the ring of all the Gardens.

The cream of New York's crime scene attended the last fight night at Garden II on 25th Street. It was 5 May 1925, an evening that dripped in schmaltz. Tiny Joe Humphries, the Michael Buffer of his day, wiped the traces of tears from his eyes, dragged down the overhead microphone and intoned with all the solemnity he could muster (which was considerable): 'Before presenting the stellar attraction in this, the final contest in our beloved home, I wish to say this marks the "crossing of the bar" for this venerable old arena that has stood the acid test these memorable years. And let us pay tribute to Tex Rickard and the other great gentlemen and sportsmen who have assembled within these hallowed portals.'

Characterising Rickard as 'a great gentleman' stretched the sinews of old Joe's irony cells. Only three years earlier Tex had to draw on every smart friend he had, from politicians to lawyers, to extract himself from messy allegations that he'd pestered young girls. Old Joe soldiered on nevertheless. You could almost hear the violins from some celestial eyrie as he wound up: 'Goodbye then, old temple, farewell to thee, oh Goddess Diana standing on your tower. Goodnight all . . . until we meet again!'

Cue thunderous reception – and on with the motley.

The closer that summer's night of '25 brought together Johnny 'The Scotch Wop' Dundee (he was born in Sciacca, Sicily, and grew up in New York as Giuseppe Carrorais until his pro career started and he made an apparent nod towards a Caledonian constituency) and Sid Terris over twelve rounds at featherweight. Sid won on points.

When the tumult subsided, and the boxers gathered together their kitbags to leave that Garden for the last time, one John F. Mullins strode into the ring in the distinctive colours of the Fighting 69th, bedecked with his war medals, and played taps. Hallelujah!

This was the height of the roaring twenties, and Rickard's reign at Garden III, although it would be brief, was about to begin. The bootleggers, criminals and various investors could hardly wait. Rickard, who'd promoted Jack Dempsey, co-opted the future New York State governor W. Averill Harriman to join his consortium of investors at the new establishment. In the hectic tempo of the decade, it took a mere 249 days to build the place on Eighth Avenue and 49th Street.

By most authoritative accounts, the first fighters to step into the new ring at Garden III were Paul Berlenbach and Jack Delaney, on Friday night, 11 December 1925. Rickard was the promoter – alongside one Vince McMahon, the grandfather of the Vince McMahon known to wrestling fans today as the owner of and sometime participant in World Wrestling Entertainment.

Berlenbach and Delaney contested Berlenbach's light-heavyweight title over the championship distance of the day, fifteen rounds, and, inevitably, not all was as it seemed. Some of the

17,575 customers who'd filed in from the speakeasies around Broadway that night no doubt imagined Delaney was Irish, a sure ticket-selling bonus in those days. Jack was, in fact, a French Canadian called Ovila Chapdelaine. So, for the purposes of commerce, the chap Delaine morphed into the chap Delaney. We can be reasonably sure Paul did not change his name to Berlenbach; most things German had little cachet after the Great War.

Whatever their real names or worth as fighting men, this is how the fight was recorded in the papers of the day: 'Floored for a count of three in the fourth and punched groggy in the sixth and seventh, Berlenbach came back in the last six rounds with a stirring rally which saved for him the title he had wrested from Mike McTigue. His margin of victory was close, however, for newspapermen at ringside gave him only seven rounds, to six for the challenger, while two were even . . . But the outstanding factor in his success was indomitable courage in the face of almost certain defeat.'

Rickard pushed that fight as the first in the new Garden. However, a disputed and little-known version exists which maintains the first fight in the new 1925 ring occurred three nights earlier, an amateur flyweight contest between Jack McDermott and Johnny Erickson. But no respected archivist has been able to confirm it took place.

What is not in dispute is Rickard died on 6 January 1929, cut down at fifty-nine after an operation for appendicitis went wrong. It was nine months before the Wall Street Crash, not a bad time to check out. He left behind a string of memorable nights. The fighters who flocked to New York in the twenties, most of them performing at the Garden, most of them paid by Rickard, included Benny Leonard, Jack Dempsey, Britain's world champions Jimmy Wilde, Ted Kid Lewis, Jack Kid Berg and Teddy Baldock, Georges Carpentier, Mickey Walker, Jack Britton, Irishman Mike McTigue, Pancho Villa, Luis Firpo, Harry Greb, Jimmy McClarnin, Jack Sharkey, Tommy Loughran and Max Schmeling.

*

As the twenties closed, America was still 'dry'. But citizens were tired of Prohibition, tired of Big Government, tired of being pushed around. The restrictions spluttered on until 1933, as the Depression wiped out jobs and hope. In the thirteen years of its life, Prohibition had given the Mob time to establish the sort of dominance it had dreamed of. The gangsters had control of the triple thrills of drinking, gambling and fighting. Like the financial crisis sweeping the world, nobody knew how, where or when it would end.

What everyone knew was the Garden was now the hub of boxing, the nation's, the world's most accessible and glamorous sports entertainment. The lights of Broadway had worked their magic yet again.

This was the state of the game when Mike Jacobs, now near fifty, aspired to step into Rickard's shoes. He'd hung in there, busking it quite often, rarely using his own money, and insinuating himself deeper into the upper reaches of the business's hierarchy. He knew every fighter, manager and promoter in the business. He didn't necessarily like them, and most of those he did business with didn't much care for him. Surely, though, he would be the new Rickard.

Not straight away. The old power structure remained in place. The Garden was not there to be taken without a fight. It was in the hands of people who loved power, influence and money just as much as Jacobs and his friends did. The expatriate Liverpudlian Jimmy Johnston was the matchmaker at the Garden, and would remain so for a few years yet. But Jacobs was a patient man. And his associates had the sort of money that brooked no argument. His friendship with Damon Runyon, in particular, would prove crucial in the years to come.

4

OWNEY MADDEN AND LUCKY JIM

A lot of desperate men resorted to earning a living with their fists when most of the legal options disappeared during the Great Depression. And, where there are desperate men, there are others more than willing to accommodate their desperation. Owen Madden was such a man.

Madden's widowed mother, an O'Neil, sailed from Liverpool to New York in 1901 with her sister, landing up in Hell's Kitchen, the West Side slum that Irish refugees from the potato famine had made their own uncoveted manor for nearly fifty years. From the moment Owen followed a year later, aged nine, to join his mother and her sister in a cold-water flat on Tenth Avenue, he traded strongly on his Irish roots.

His friends called him Owney. While he was growing up in the company of Arnie Rothstein, 'Lucky' Luciano, Jack 'Legs' Diamond and Dutch Schultz, none of them choirboys, Madden, born in Leeds but made for New York, was also known more chillingly as The Killer. He was a top-flight thug, a graduate of the feared Gopher Gang who earned his reputation wielding a gun as if he were in a Wild West show. After his uncontrollable temper got the better of him once too often, he spent nine years in prison for murder.

When Madden got out of Sing Sing, in 1923, there to meet him at the gates was Joe Gould, sitting at the wheel of a fancy Packard, alongside a convicted murderer called Arthur Bieler. Gould, a small-time boxing hustler, had been instructed to collect Madden on behalf of Schultz, one of New York's premier liquor salesmen. Madden got in and Gould handed him a beer. 'This is what Dutch Schultz puts out,' Bieler told him. 'If you play your cards right, you can get in on the act. Dutch don't like no fuckin' freelance operators. You would do well to do what he says.'

Madden didn't much care for the quality of the spiel, or the beer – and he spat the latter all over Bieler's pinstriped suit. Owney determined he would open his own brewery. He told Bieler to tell that to Schultz. To start out as a bona fide bootlegger, though, he had to get rid of Eugene O'Hare, a fellow Irish American Schultz had installed on Madden's patch after he was sent up the river. Within weeks, O'Hare's dead body was found on an empty scrap of land on the Lower West Side.

Gould, meanwhile, was soon to play a role quite a deal more important than driving a Packard to Sing Sing for Dutch Schultz. In the course of boxing history he would not have envisaged a role much grander than that – but for his association with Owney Madden. Gould was the manager of James J. Braddock; Owen had a large piece of Max Baer. And one day they would collide.

Braddock, born in Hell's Kitchen, raised in Jersey, was just another 'Irish' fighter as the thirties got under way, a good one among thousands, but not exceptional. He fought often – and often he fought with cracked ribs, sore knuckles and not much food in his belly. But, as was the norm, he had to do business with people who put a better gloss on their reputations than was deserved.

Gould was indisputably one of those chameleons. Joe was close to Madden – and that was not like being next to cleanliness. With Prohibition came endless business opportunities, chiefly at the shakedown end of the retailing gig. You could sell Madden's booze, if you signed up for Madden's goons to 'protect' your premises and your back. Madden worked for Schultz only as long as it took him to stamp his own authority on his patch, Chelsea. With money from strike-breaking and bootlegging, Madden opened the Cotton Club – previously owned by the world heavyweight champion Jack Johnson when patrons knew it as the Club DeLuxe. (Years later, the establishment and its owner achieved immortality of sorts when Bob Hoskins played Madden in Francis Ford Coppola's eponymous movie of the infamous hangout.) Owney also had a stake in the Stork. If you were looking for crime and criminals, these were the places to be seen.

Madden thrived in Chelsea. His brewery occupied a building so obvious – on the Lower West Side, on Tenth Avenue between 26th and 27th Streets – it must have been difficult for policemen of the day to walk past without putting their hands over their eyes. His beer, called Madden's Number 1, was, by all accounts, not bad, and he served it in pints. Damon Runyon asked him once if he were not stretching the patience of the law by putting his name on the bottle. Madden said it was the most popular beer in New York and he was 'dead proud' to have his name on the label.

According to Jimmy Breslin, Madden took Runyon to one of his fancy apartments one night, a penthouse at the top of the London Terrace block on 23rd and Ninth Avenue. From the rooftop, they could see his brewery. Staying in the apartment were Ray Arcel, who would go on to become one of boxing's most revered trainers, and his fighter, Charley 'Phil' Rosenberg. Madden was now entrenched in the boxing business and had arranged for Rosenberg to fight Eddie 'Cannonball' Martin for the world bantamweight title. Rosenberg was in his apartment because Madden wanted to keep an eye on the challenger's diet. To that end, he also installed in the swanky apartment Charley Phil's cook, his mother.

Come fight night, Madden bet $1,000 – on Martin, because he

didn't think Phil had been eating right. Charley Phil cut the champ to pieces. Owney was livid.

Madden is sometimes overlooked in the history of gangsterism's grip on boxing, but it was during the turbulent thirties that he was at the height of his dubious powers. Along with Schultz and Vincent 'Mad Dog' Coll (whom he'd later help kill), Owney had a piece of the world heavyweight champion Primo Carnera – even though, as far as the National Boxing Association was concerned, the champ's managers were Louis Soreci, Billy Duffy and Walter Friedman. Next to Joe 'The Human Punch Bag' Grim, who was knocked down at least eighty times and won maybe four of 113 verified contests in the thirteen years leading up to the Great War, Carnera was the most pathetic figure in all of boxing. At the end, he owned less than 10 per cent of himself.

Madden was near the heartbeat of the sick beast Boxing throughout the Depression. At one point, he controlled the bantamweight and light-heavyweight title holders, as well as four of the five heavyweight champions between 1930 and 1937. Max Baer was the only fighter to hold out against him, publicly at least. It was always rumoured Madden had a piece of him too, through Duffy and an underworld tough called Big Frenchie Demange. Gould was now on the inside, with a so-so heavyweight and connections to all the people who controlled the title.

Owney, it is said, also owned a slice of Braddock. And you won't see that in the schlock movie *Cinderella Man*, which depicts Braddock as an innocent victim of his times and calling, oblivious to the deals Gould was doing with the psychopath Madden.

Braddock's story was a good one. Devastated by the Wall Street Crash, he famously rose from the breadline and occasional work on the Jersey docks and, against all the odds, got a shot at the awesome puncher Baer.

Gould expended a lot of energy in putting the hitherto down-and-out Braddock in the limelight, introduced him to influential friends of all stripes. Later, he would pull off one of the cheekiest scams in boxing history, but first, they had to get their hands on the title.

Gould was a hypemeister to match it with the best of his day. He was not fazed by threats or demands, even from the men in pulled-down hats. He knew many of them as friends. He had Madden on board. He knew Carbo, Palermo, Costello and J. Edgar Hoover, the head of the FBI. This was some guardian angel. While selling Braddock to a sceptical audience, he maintained a classic front, retaining an office he could not afford, making promises he wasn't sure he could keep.

Baer, known here and there as 'The Livermore Larruper', was an awesome hitter. He'd killed one man in the ring and gave a beating to another who died later, tragedies which took the edge off his aggressive instincts, and made a light-hearted man occasionally sad. Might it also have contributed to what has been described down the years with ever increasing conviction, even by those who were not there, as the biggest upset in heavyweight history?

Max was more than a stone heavier than the challenger, and should have murdered Braddock, in the nicest possible way, that night in the Madison Square Garden Bowl, an offshoot of the MSG, in nearby Long Island. But he hadn't trained properly and there was a lot of the clown in Max, who loved the high life. He took James too lightly. Braddock found something deep down he always knew was there. It came together for him over the fifteen rounds in his career that mattered more than any other, and he beat Max. His opponent wished Jim well, saying he hoped he valued the title more than he had done, a gnomic reference perhaps to the pressures brought to bear behind the scenes by Madden.

Lucky Jim was now not only the world heavyweight champion, he was the unofficial king of New York, with the run of the clubs and bars up and down Broadway and its environs. Runyon was the first to call him the Cinderella Man, and it stuck, another of boxing's enduring fairy tales. What happened before and after that historic upset is less well chronicled.

In the engrossing *Official and Confidential: The Secret Life of J. Edgar Hoover*, Anthony Summers offers this real-life glimpse of Braddock's movements away from the ring:

> The mob bosses had been well placed to find out about Edgar's compromising secret, and at a significant time and place. It was on New Year's Eve 1936, after dinner at the Stork Club, that Edgar was seen by two of Walter Winchell's guests holding hands with his lover, Clyde [Tolson]. At the Stork, where he was a regular, Edgar was immensely vulnerable to observation by mobsters. The heavyweight champion Jim Braddock, who also dined with Edgar and Clyde that evening, was controlled by [Frank] Costello's associate Owney Madden. Winchell, as compulsive a gossip in private as he was in his column, constantly cultivated Costello. Sherman Billingsley, the former bootlegger who ran the Stork, reportedly installed two-way mirrors in the toilets and hidden microphones at tables used by celebrities. Billingsley was a pawn of Frank Costello's, and Costello was said to be the club's real owner. He would have had no compunction about persecuting Edgar, and he loathed homosexuals.

This was not a fairy tale. This was the Cinderella Man out on the town with the head of the FBI and his boyfriend, surrounded by an array of unsavoury types, as well as the biggest and most self-important windbag in town, Walter Winchell. Even back then, Frankie Carbo, who would go on to be the main man in the fifties, was in on the take, a regular at the Cotton Club and the Stork. It is said he was answerable only to Costello, and had regular quiet talks with Mike Jacobs.

So, whatever the halo that film-makers might have deposited over their innocent heads, Joe Gould and James J. Braddock moved with ease through all parts of Gotham, from the Garden to Lindy's, the Cotton Club and on to the Stork. Sometimes, they'd stop in at Dempsey's and hang out with Jack. Dempsey's got a reputation over the years as a drop-off place for the Mob's bag money, a sort of gangsters' post office, but none of this ever rubbed off on the proprietor – who also played a background part in Max Baer's career.

Was Baer–Braddock a fix? There is no evidence. But Baer was the only heavyweight of the era not to go openly with Madden. Owney wanted him out of the picture. And it suited him and others to have Joe Gould's fighter as the heavyweight champion of the world, because he could be manipulated more easily, through Gould, his one-time point man to Dutch Schultz. It is inconceivable that Max took a dive – and what was to follow does not constitute a case for the prosecution. But, whatever the reality, it all fell neatly into place for Gould and his Cinderella Man.

The fight with Baer was, in truth, a one-off for Braddock. It has been dressed up over time as the greatest upset in the history of heavyweight boxing up to that point when, in fact, it was an honest, workmanlike performance. The fight itself was dull. When Max threw punches, they rocked Jim; trouble was, he hardly threw any – and Braddock just pecked away and survived. Every fighter has a story, often a good one, and this was Jim's. But the reality was Braddock fought the right fight on the right night against a champion who clowned about once too often.

Max, who might have had a long and lucrative reign, was consigned to boxing's second division without fuss. Had he gone with Madden, that might not have happened. Instead, Joe Louis – now part of the action at the summit of the fight game thanks to the encouragement of Mike Jacobs – gave him a memorable four-round beating in his next fight. Thereafter, over the next two years, Max belted out a decent living against other hopefuls, twenty-nine of them, but he would never get another shot at the title, even though he deserved one.

There was a curious symmetry to Baer's career, which he wound up in 1938: Max fought a total of 110 professional rounds and scored 110 knock-downs. The *Ring* magazine rated him at 22 among the 100 greatest punchers of all time.

Once he stopped trading on being Jewish (Goebbels banned his first movie in Nazi Germany 'not because I was Jewish, but because I beat Schmeling'), he made the most of his looks to earn a good living as an actor. Max was a bright-eyed, smiling presence in a succession of forgettable movies and TV shows, all the way up to 1959. His son, Max Jr, found screen fame as Jethro in the sixties TV hit, *The Beverly Hillbillies*, and he was livid when *Cinderella Man* portrayed Baer as a mean, unfeeling fighter. He was anything but. As far as his family and friends were concerned, Max was a rush of mountain air in a sewer.

For Braddock, life took a different turn. While the boxing fraternity and the general public were buzzing with the enormity of his achievement against the fearsome Baer, the game's machinery was grinding away to ensure this well-placed champion and his manager did not go short once he faced the inevitable: a big-money showdown with the unbeaten Louis and the loss of his title.

However good and fleeting a story Braddock's was, everyone at the time knew the best heavyweight boxer in the world was Joe Louis. Mike Jacobs knew, the kids stealing nickels from the phone box knew, the President of the United States knew. Jimmy Johnston wasn't so sure. And the reason was Joe Louis was black.

Johnston's reign as matchmaker at the Garden was, predictably, troublingly racist. The Garden, advised by Johnston, reflected the wider view in the business that fight fans wanted a white heavyweight champion. It was a prejudice that was to cost Jimmy the ride of his life.

In a single phone conversation with Joe's manager, John Roxborough, Johnston properly gave the game away.

Roxborough rang Johnston to see if he could find a place for Louis on a Garden bill, after Joe's string of impressive wins in the Midwest, including two over the well-regarded Lee Ramage.

'If he comes here,' Johnston told Roxborough, 'he'll be expected to lose a few. I don't care if he's knocked out Ramage. He's still coloured . . . Don't you have any white boys out there?'

What Johnston couldn't know over the phone was Roxborough was black – as black as Joe, who would go on to be the biggest box-office hit in boxing since Jack Dempsey and until Muhammad Ali. And Johnston missed out on signing Louis because his manager didn't much like Johnston's attitude. Such biological determinism as the Garden matchmaker's would not have looked out of place in the bierkellers of Munich at that time.

Louis, contrary to the heavily massaged perceptions of his day, suffered at the hands of all types and races, including his own. He made a lot of money – for other people, mainly. He had a black manager, a Jewish promoter and, waiting up the road, a Jewish-managed German opponent, Hitler's pet Max Schmeling, who would spend the rest of his life vigorously detaching himself from the prevailing philosophy of the Third Reich and all the sins they visited upon Hebrews everywhere.

This is the story of what happened to Joe Louis, and Mike Jacobs's part in it.

A couple of weeks after Baer–Braddock, Joe was scheduled to beat Primo Carnera – from whom Baer had won the title in 1934 – and Louis duly did what he had to do. But, the preliminaries out of the way, what might have seemed a natural match between the finest heavyweight of that part of the century, Louis, and the new champion, a man blessed (or otherwise) to share the same occupation, Braddock, did not pan out as expected – except by those in the know. And nobody was more in the know than Mike Jacobs.

Jacobs was a strange fish. He had no worries referring to all dark-skinned fighters as schwartzes – but at the same time he would give them as much work as they could handle. His only prejudice was against an unimpressive bank balance. He knew Louis was good business. In time, he would also come to be proud of their friendship, not to mention their business relationship.

Roxborough and his partner, Julian Black, an old numbers racketeer, were as attuned to the moneymaking possibilities of their boxer as was Jacobs. They knew that, while Johnston was a roadblock, Jacobs was a conduit. What they had to overcome, however, was the ingrained resistance of white America to recognising a black man as the best heavyweight in the world. Roxborough needed no history lessons on this subject.

The reason Louis was having trouble making his way to the top of the heavyweight pile was the legacy left by the last black champion, Jack Johnson. Jack gave the White Hope, James J. Jeffries, such a hiding in Reno, Nevada, on 4 July 1910, that few present would ever forget it.

Tex Rickard, who promoted Johnson–Jeffries, was anxious to give the bout a hint of respectability, and even asked President Taft to referee. The president said he was busy. Johnson was indisputably the best heavyweight of his time, the champion of the world since he'd ripped the title away from Tommy Burns in Sydney on Boxing Day, 1908. They'd dragged old, white James J. off his alfalfa farm in Ohio, five years into his retirement, to 'put down the uppity nigger', this refugee from the Chitlin' Circuit, who'd once had to be content to box his black brothers, among them the similarly gifted Sam McVey and Sam Langford, and then had left them far behind too.

But Jim, thirty-five, was too slow and Jack was too good. Johnson, who self-mockingly described himself as 'the brunette in a blond town', tormented Jeffries before putting him out of his misery in the fifteenth round in front of an audience of cowboys, hookers and thieves.

For a taste of the evil of the times, Johnson himself is as reliable a source as many. He says this in his autobiography:

> More than 25,000 people had gathered to watch the fight and, as I looked about me and scanned that sea of white faces, I felt the auspiciousness of the occasion. There were few men of my own race among the spectators. I realized that my victory in this event meant more than on any previous occasion. It wasn't just the championship that was at stake: it was my own honor, and in a degree the honor of my own race. The 'White Hope' had failed.

Hysteria greeted Jack's deed. Blacks were murdered in race riots across America – nobody is sure how many – lynched and humiliated by the Klan and other white supremacists. The colour bar went up just as Jack was trying to tear it down, albeit for his own purposes.

As the acclaimed American film-maker Ken Burns said in his 2006 documentary *Unforgivable Blackness: The Rise and Fall of Jack Johnson*: 'Johnson in many ways is an embodiment of the African American struggle to be truly free in this country – economically, socially and politically. He absolutely refused to play by the rules set by the white establishment, or even those of the black community. In that sense, he fought for freedom not just as a black man, but as an individual.'

His people even wrote a spiritual about him:

> Amazin' Grace, how sweet it sounds,
> Jack Johnson knock Jim Jeffries down.
> Jim Jeffries jump up an' hit Jack on the chin,

An' then Jack knock him down agin.
The Yankees hold the play,
The white man pulls the trigger,
But it make no difference what the white man say,
The world champion's still a nigger!

And white did not want black ever to have even a chance of being champion again. Some black people agreed. Pastors and kindly old community leaders preached silence. Booker T. Washington, the conservative black intellectual fountainhead of his day, reminded his brothers and sisters, 'No one can do so much injury to the Negro race as the Negro himself.' Angry blacks called that just another day of slavery. And number-one black of the day, Jack Johnson, said it loudest and longest.

W.E.B. Du Bois, the radical alternative to Washington, said: 'Jack Johnson has out-sparred an Irishman. He did it with little brutality, the utmost fairness and great good nature. He did not knock his opponent senseless. Apparently he did not even try. Neither he nor his race invented prizefighting or particularly like it. Why then this thrill of national disgust? Because Johnson is Black.'

Du Bois was shaky on the boxing facts, and on the money with the social consequences. It would be well into the Great Depression before another black man would be in position to challenge the white man's stolen supremacy as the best heavyweight in the world.

Roxborough knew his history, all right. That's why, as a businessman, he would do whatever it took to get Joe his shot. It might not seem that savoury from a distance, but he had to deal with the prevailing morals and attitudes of the day.

It was his good fortune to be doing business with Jacobs at just the time Uncle Mike was rising to the top of his profession, ready to make a move on the big job at the Garden. Johnston's job.

Having made no progress with Johnston, Roxborough got to

work on Jacobs. Roxy went out of his way to assure the almost totally white fight-writing fraternity that Joe Louis would not be photographed with white women, wouldn't be seen alone in nightclubs, wouldn't have an easy or fixed fight, wouldn't humiliate a beaten foe by standing over him in triumph, would not 'showboat' in any way, but would be a clean-living young athlete of which America, all of America, could be proud.

Joe went along with most of it – apart from the 'clean-living' bit; he chased women, white or black, voraciously.

But they did a deal. Jacobs did the deal of his life; Joe did the only deal on offer. He wasn't complaining. All he wanted to do was fight. He didn't read the contracts, he just listened to his handlers, got fit and knocked out anyone they put in front of him. To Joe, it could not be simpler. Eventually, of course, it would become so complex it would wreck his life. For now, though, he was a young, black fighting genius rushing towards his destiny.

As the great adventure was getting under way in 1935, then, Jacobs was the man in control of Joe and the title. He did as he wanted with both, and he was not without friends, naturally. He had been co-opted on to what was to become the Twentieth Century Sporting Club and it had as its prime movers the pre-eminent media dictator of his day, William Randolph Hearst, plus Ed Frayne, the sports editor of the *American*, Bill Farnsworth, sports editor of the *Journal*, and the legend himself, Runyon, Hearst's favourite and most eloquent mouthpiece. It was a powerful team, and they had their eyes on the Garden. Johnston and his bosses didn't stand a chance. Joe was on his way. Just about.

After Jacobs got Roxborough to sign Louis up to the Twentieth Century Sporting Club, to promise white America that black America would not visit another Jack Johnson upon their precious heads, everything fell into place. It seemed so easy now for the scuffling Joe Gould to make lots of money. He did not let his boy

Braddock down. Once Jim had the championship, Gould came into his own. He engineered a quick autobiography, *Braddock: Relief to Royalty*, and a flimsy legend was born. That was just the start of it.

When Joe knocked the stuffing out of the Ambling Alp, Primo Carnera, up the road at the Yankee Stadium (attracting 64,000 customers, twice the number who filled the Garden for Baer–Braddock), it seemed the clamour for Louis–Braddock would not be denied for long. It was put on hold – but for reasons that would not become immediately obvious. This was going to be a slow burn, kindled by the Hearst newspapers, which now had links to the Garden through the enterprising Runyon and his pals . . .

The lineage of boxing's most prized championship up to that point was, generally, a distinguished one: Sullivan, who ruled with his knuckles for several years, was legitimately a world champion, by general consensus, from his bare-fisted win over Jake Kilrain in 1889 until he lost with gloves to James J. Corbett in 1892. Corbett lasted until '97; Bob Fitzsimmons, '97 to '99; James J. Jeffries, '99 to 1905; Marvin Hart, '05 to '06; Tommy Burns, '06 to '08; Jack Johnson, '08 to '15; Jess Willard, '15 to '19; Jack Dempsey, '19 to '26; Gene Tunney, '26 to '28 (the title was in abeyance for two years after Tunney retired in 1928); Max Schmeling, '30 to '32; Jack Sharkey, '32 to '33; Primo Carnera, '33 to '34; and Baer, who held it a day short of a year.

That made it fourteen champions in forty-three years, almost a royal succession. What was to follow would do much to eat away at the credibility of professional boxing. After Braddock and all the way up to late 2006, just seventy-one years, the title changed hands a bewildering ninety-three times. What started as the biggest prize in sport became a very grubby enterprise which then somehow managed to slide even further.

At least Lucky Jim's recognition meant something. He was all over the papers. People shook his hand in the street, bought him a drink in the clubs. He fitted the picture. He'd risen from bum to hero in a twinkling. If he could do it, anyone could. America needed Jim Braddock badly in 1935, as the Depression gnawed away at the fibre of its being. To the guys growing weaker by the day at the soup kitchen, or riding the rails, Jim was one of them. Just like Dempsey had been. It didn't matter that he wasn't a great fighter. He fought great on one night. He sustained hope.

But Braddock too would be dispensed with soon enough. Asked how long the new champ might remain in office, Max Baer told reporters, 'Until he fights somebody else.'

5

SETTING UP JOE

'Somebody else' was at ringside the night Lucky Jim won the title. He even dropped off to sleep between rounds, so dull an affair was it. But then Joe Louis was always an unusually relaxed and patient man.

Not a lot moved him, save a big left hook when he wasn't watching, or a disapproving glance from his first wife, Marva, after he'd been caught out yet again. Joe, for all his physical and athletic strength, was a prisoner of forces beyond his control. He sometimes would say, in expectation and hope, 'God is on our side.' A simple phrase, which became something of a slogan for GIs during World War II, it not only described Joe's fatalism but identified a peculiar strand of benign righteousness in America. It was a sentiment that would be the bedrock of his career. It made him acceptable, a good American. That night back in '35, meanwhile, Joe was content to let his Lord anoint James J. Braddock the winner – and make him a prime candidate for the chop further up the road.

Paul Gallico, a sports writer possessed of both wit and witlessness, once described Joe's quietness as 'sly servility'. Gallico, the son of Italian immigrants, a born New Yorker to the tips of his fancy shoes, a man of no little intelligence whose father wanted him to be a concert pianist, should have known better. But he still did not have the nous or instincts needed to make the intellectual leap away from the herd. He was, Robert Lipsyte generously said many years later, 'a man of his times'. As Lipsyte acknowledges, though, that always was the lamest of excuses for clever racism.

Gallico, an evocative if reflexive writer, was a friend of Runyon's. He was a friend of anyone in sport, in fact, as long as they'd support his commercial ventures. And there were many of those. His critical integrity was easily bought. Just like Damon's. They belonged to a distant phenomenon, the sometime entrepreneur moonlighting as commentator and social wit. Gallico and Runyon covered many cultural and sporting waterfronts – then went to their typewriters to shape the market for their enterprise. They described the mood of a New York ripe for such description, and their perspective was, like the skin that housed their bones, unequivocally white.

Lipsyte, a childhood fan of Gallico's whose affection waned as he grew into the sports-writing gig himself, sees the fingerprints of prejudice all over the man's work, lyrical and lovely as it sometimes is. At ringside, for instance, Gallico looked up from his free seat, his telegraphist by his side, his cigar tucked away somewhere in his silk-lined jacket, and saw not just two fighters trying to earn an honest (or dishonest) dollar but, on occasions, 'the colored brother', as he liked to dress it up. And 'the colored brother' might triumph – if he were in against a white fighter – not necessarily because of his skill or courage alone but because, he was 'not nearly so sensitive to pain as his white brother. He has a thick, hard skull and good hands.'

That was the climate in which Joe had to make his mark. He had two things going for him: his left and his right.

Joe's star was hanging high in 1935 and James J. Braddock, the luckiest Jim that ever got up in the morning, would hitch himself to it before long – or rather his supposedly avuncular manager would do it for him. Gould knew he had to move deftly to maximise his earnings before 'somebody else' came along and put Jim out of the picture.

Louis needed the title, but he was not going to earn a night in the Garden straightaway. To get his shot at the Irish patsy who called himself champ, and to build his already formidable reputation, he was encouraged to bide his time with easy wins, to go with the nineteen he had registered since turning professional in 1934. He'd already detonated the myth of the Ambling Alp, Carnera. In quick order and with a coldly simple left-jab, right-cross efficiency, he went on to dispose of King Levinsky in Chicago in August, Baer the following month back in the Bronx, Paulino Uzcudun before Christmas, Charley Retzlaff in January . . . and the rumbling grew for something more meaningful from the prodigy. The strategy was working.

Then a bombshell landed. Or rather, an 8–1 German underdog landed. The contest between Joe Louis and Max Schmeling at Yankee Stadium in June 1936, a year after Braddock won the title, shocked not just boxing fans and the wider world, but the guys who made things happen. This wasn't scheduled. And it was on the level. The Brown Bomber, a 10–1 on favourite (as Baer had been against Braddock), was blitzkrieged off the boxing landscape. He was outpunched and outboxed by a decent opponent who, on the biggest night of his life, got it right. Just like Braddock. For a sport that was supposed to be riddled with arranged results, the no-hopers were doing OK.

Maximillian Adolph Otto Siegfried Schmeling, the Black Uhlan, so called because of the heavily dark complexion common to people from that part of Germany, shared a name with Hitler but not, according to all reliable testimony, his thirst for genocide. Nevertheless, through expediency or inclination, Schmeling's reluctance to endorse his Führer was not always as enthusiastic at the time as it has become in the revisiting of the story.

What is clear is, as a professional fighter, he put the chance to beat the most feared heavyweight in the business above the inconvenience of representing a regime dedicated to wiping out the Jews and subjugating all other peoples with no obvious links to the Aryan race. Apologists at the time, such as leader writers on the *Daily Mail* and *The Times* of London, looked the other way. Revisionists later, such as Gallico, stressed how Max never had Hitler around for supper. But, as David Margolick notes in his must-read study on the fights between Louis and Schmeling, Max did vote along with 48,799,268 of his compatriots for the annexation of Austria in April 1938. It is hard to accept that Schmeling, a bright man who would later become a wealthy franchise holder in Germany for the most American of products, Coca-Cola, was a complete innocent.

Nor was he a mug as a fighter. In the twelfth round of their first fight, Schmeling hauled down Joe's star. Joe, who'd grown complacent in the glow of adulation, hadn't trained well. Max had. The German belted Louis senseless and left him shaking his head as he sat bewildered, hurt and friendless on the canvas. Max deserved his win; he'd done it on the night, and Joe had let himself down. That's sport. It should neither have vindicated Nazism nor demeaned Louis. But it did both those things in many people's eyes, for reasons that now seem obvious, but did not seem so at the time.

The picture of Joe Louis sitting on the canvas staring into the middle distance, with the German standing over him, was one America did not want to look at. It was the worst sort of news for the other main players, too. Especially for Joe Gould. Initially, he'd wanted Braddock to fight Louis sooner rather than later because he knew that was the one big go he and his fighter would ever have at making a pile. The apparently foolproof plan that emerged after the Braddock–Baer fight was to build Louis and mothball Braddock. Nobody counted on Schmeling beating Louis.

With Louis's invincibility punctured, they had to put the Brown Bomber back together again. Braddock still had not laced up since the Baer fight; Joe had boxed six times.

After Schmeling, Louis would have another seven bouts before he got in the ring with the unemployed champion Braddock. When the dust settled many years later, Gould could reflect on a job well done. This was to be the long-term pay-off, one worth waiting for, according to Braddock's cagey, dodgy manager – even though his client, at the first time of asking, would lose his title and give up the cachet, notionally at least, of being the best heavyweight on the scene.

Away from the headlines, Jim's manager played a blinder. He saw in Louis the most malleable, marketable of champions – and he wanted a piece of him. Braddock, his buddy, was no more than the means to get to Louis. This was Mob strategy – skimming, as it's known in gambling and liquor circles, taking the vigorish off the top. Creaming it. What Joe's connections and others then did to the man who was 'a credit to his race' would be regarded as cruel in any other undertaking but professional boxing.

When the unaffected and trusting young man from Detroit turned twenty-one in 1935, he signed over half his gross earnings for the following ten years to his first manager, Julian Black; his other manager, Roxborough, claimed a quarter 'for an indefinite period'; his trainer, Jack 'Chappie' Blackburn, a convicted felon but a man the champ considered his one true friend in boxing, took his wages from the quarter that was left for Joe. Then the taxman went to work on Louis. It was a salami slice. From the day he entered the gym back in Detroit, Joe never stood a chance.

Initially, Madison Square Garden wanted Braddock to defend against Schmeling in June of 1937. But Mike Jacobs wasn't so keen. Not yet. He persuaded Gould to stall on any offers from the German, to keep his fighter for Louis later. Gould agreed – but he wanted the deal sugared. According to some accounts, about this time Roxborough was picked up by some persuasive gentlemen and taken to see Gould. The champ's best friend told Roxborough he didn't reckon Jim could beat Joe if he hit him until the sun came up. But, to make sure there was no upset, Gould wanted 50 per cent of Louis. Roxborough, sweating hard in front of the interview panel, held out. 'I can't do that, Joe,' he said, with good mathematical reasoning. 'If you want to do a deal, you gotta talk to Mike.'

Gould was outraged. He went to the Garden and talked to Mike. It was a conversation that would leave an enduring scar on Joe Louis.

'Mike, I'm a reasonable man,' said Gould. 'I'll settle for 10 per cent – or there's no fight.'

Jacobs, who'd been in the game too long and had muscle of his own, laughed in Gould's face. 'Joe, you're crazy. There ain't no 10 per cent to give ya! Didn't Roxy tell ya that?'

He hadn't. The boys had already carved Joe Louis to pieces.

'We need something, Mike. We need something. You got it. We need it.'

'Tell ya what, Joe. We all need this fight, so I'll do this for ya: you can have one dollar in every ten Uncle Mike earns with the title until some schmuck gets lucky and knocks the schwartze over. Whaddya say? Joe Louis is gonna be around a long time, Joe. And so is Uncle Mike. We're all gonna earn some serious moolah here.'

Gould did his sums as quickly as if standing in front of the bookies at Santa Ana.

'Done.'

So Gould was set for a tenth of Jacobs's profits from every world heavyweight title he would promote for the next ten years. That, effectively, meant every fight Joe Louis had under Jacobs's promotion with the Twentieth Century Sporting Club. It also meant the new champ's purse would be shaved accordingly. Mike Jacobs was not about to give Joe Gould 10 per cent of Mike Jacobs; but he would give him 10 per cent of Louis.

After the deal was done – behind Louis's back – the fighter would go to work to make everyone rich. First, though, Gould had to get Braddock out of his fight with Schmeling.

Germany in '37 was an arrogant, menacing place. Hitler had held his Olympics the year before and he had in Schmeling a high-profile heavyweight with whom to peddle the message of Aryan supremacy.

There is a story, first written in 1950 by Budd Schulberg, that Joe Gould told Germany's propaganda minister Josef Goebbels in uncomplicated terms what he thought of Hitler's wish that Braddock defend his title against Schmeling in Germany.

The telephone exchange is said to have finished with Gould informing Goebbels that $500,000 up front and an American referee would not be enough to clinch the fight.

'The third point,' Gould said matter-of-factly, 'is that you get Hitler to stop kicking the Jews around. Unless he gives them back full citizenship and property rights, you know what you and Max can do with your fight.'

Gould the saint and wit? It is a departure from everything we know about Gould to regard him as a moralist first and a businessman second. Besides, who'd want to take the title to Germany?

David Margolick pointed to the pages of the *New York Daily News* of 22 March 1935, to capture a different picture of the times. If Gould was reluctant to do business with the Nazis, one fellow American Jew prominent in the boxing game was not. The image decimated the weasel words written then and since about Schmeling and those around him. He had just knocked out Steve Hamas in Hamburg and, in keeping with the protocol of Germany's totalitarian diktats, the winner stretched out his strong right arm in saluting his Führer. It was an increasingly common sight – and would be repeated shamefully by England's footballers in 1938 – but such displays of obsequiousness raised indignation beyond the boundaries of the Third Reich. And this picture was different. Standing near Schmeling, hand also raised in tame homage to a totalitarian lunatic, and prevented from full extension of the fingers only by the cigar in his grip, was the German boxer's Jewish manager, Joe 'Yussel' Jacobs. The *Daily News* headline over the picture read: 'WHEN YUSSEL WENT NAZI'. He wasn't alone.

Yussel wanted Braddock for his man, and Schmeling even came to America to sign a contract to fight the champ. But Gould wanted no part of Schmeling; he wanted Louis for the champion. This seemed odd from a boxing perspective. Why would Braddock want to test against himself the murderous punching of Louis, rather than the less threatening work of the aged Schmeling? The answer, of course, was Gould and Braddock couldn't lose against Louis. They had 10 per cent of him, whatever the result.

While Schmeling posed for pictures and spent time at the Garden, hustling up interest in a non-event, Braddock and Louis were in training for their title fight. In May, a federal judge named (believe it or not) Guy L. Fake ruled the Garden could not force Braddock to fight Schmeling.

Meanwhile, although he was the centrepiece of the sport and generated the interest that kept fight writers busy, Joe Louis struggled to convince some of them he could bounce back from the Schmeling defeat. He was fat and lazy, according to many of these sages, and disposable. They were fine to his face, of course, but patronising – and they would not kiss his big black ass, even for another two-buck bribe from the promoters.

While they willed him to lose against Braddock, even their ingrained prejudices could not drown out common sense and 80 per cent of them in the end plumped for the challenger when they met at Chicago's Comiskey Park on 22 June 1937, a year after Schmeling had creamed Louis and two years after Braddock had beaten Baer. Jack Dempsey, who never put his title on the line against a black man, picked Braddock, as ordinary a champion as the division ever had.

The fight wasn't a classic, but it didn't lack for drama. Louis hit the floor in the first, from an uppercut by the thirty-two-year-old champion. Cinderella Man, it seems, did not want to go home before his carriage turned into a pumpkin. It was his last half glimpse of the prize. What punters didn't know was that Jim had a dead arm, pumped up by drugs to get him into the ring but anaethetised even further by Joe's constant battering. His left dropped lower and lower, exposing his chin to the Bomber's killer right cross. Soon enough, the fight swung the other way and, by the sixth, Braddock was spent and razored across his weathered face, but, as Gould reached for the white towel of surrender, the champion cautioned him that such an act of betrayal would be the last between them. And they'd been together eleven years, so Jim knew what he was talking about. Whatever the champion's bravery, Louis continued his clinical carving and, within two rounds, he'd dispatched the old man. He'd done it. Joe was the first black champion since Jack Johnson gave up the title in 1915 in dubious circumstances to the leering white behemoth, Jess Willard, on a sweltering afternoon in Havana.

Joe Louis, the new heavyweight king, took his mantle in the quiet and humble way his friends associated with his every movement and utterance. The ghost of Jack Johnson had been laid to rest, to the relief of white America. There were no complaints from the loser's corner.

Any right-thinking person would regard the secret deal Gould did with Jacobs as a hangover from slave times. Such was the story of Joe's life. He was exploited from the moment he taped up until the night Rocky Marciano clattered him through the ropes at Madison Square Garden for the last time, in 1951. By that stage, he'd sold another pound or three of his own flesh to the Mob. Just like Jack Johnson told him he would . . .

Johnson was there in Chicago to see Joe succeed him. Indeed, to the incredulity of all, Jack was still a licensed boxer. A year after

Louis's win over Braddock, the Galveston Giant, slightly stooped now at sixty, got into a ring with one Walter Price in Boston and was knocked out in seven rounds.

Virtually nothing is known of Price – age, nationality, where he was born, how he died – apart from the fact he had four fights in his entire professional career. The first three were in 1925, two wins and a loss against fellow novices around Massachusetts and Maine. His fourth and final bout was thirteen years later, when he beat the pension-aged illusion of a genuine ring great.

Remarkably, sadly, Johnson carried on selling himself for several years after losing to Price, dressing up in loincloth and spear as part of a travelling circus, even sparring with old opponents from decades earlier, black fighters who'd also found the bottom of the barrel. It was a wretched decline, no less shocking for its inevitability. The boxing writer Bert Sugar remembers seeing such a show by Johnson on a schoolboy visit to New York in the forties. 'Real sad,' is how he describes it. This was white America's revenge. Johnson's 'golden smile' had driven racists to distraction when he beat James J. Jeffries in 1910 on America's most treasured day, the fourth of July. He would pay for that for the rest of his life.

Joe Louis's first defence of the heavyweight title once owned by Johnson was against the Welshman Tommy Farr, who ran him mightily close in front of 32,000 fans at Yankee Stadium that August. Farr had been lined up to fight Schmeling earlier in the year in London for what the Germans and the *Daily Mail* were happy to call the real world championship. This was farcical. Schmeling had a legitimate claim to challenge Louis for the title, given he'd won so comprehensively two years earlier, but to ignore the champion's right to be considered the linear king of the world after he'd beaten the incumbent Braddock was crass. It also exposed some craven instincts among members of the British boxing and media establishment who were willing to go along with the Nazi hysteria surrounding their Max. They wanted a

Farr–Schmeling fight in London as much as Hitler did. It might well have been that patriotism and greed played a bigger part than ideology in their meek acceptance of Goebbels's entreaties, but it was no more morally convincing for that.

Mike Jacobs, however, outflanked them all. He stole Farr from under their sneering noses, brought him to America and put him in with Louis. He deserved the fight, no question; he'd outpointed Baer over twelve rounds in London in April, then knocked out the good German Walter Neusel in three rounds two months later.

The Welshman, awkward, determined and well schooled in the orthodox English way, gave Joe all the trouble he could handle over fifteen rounds at Yankee Stadium on 30 August 1937, although no nostalgia-addled rewriting of history should persuade readers that Tommy deserved to win.

Not even the *Evening Standard*, as bellicose a British flag-waver as any, saw it that way, though their correspondent, Ben Bennison, did his best. 'No fighter within my long experience,' he reported from ringside, 'has fought a braver fight for the heavyweight championship of the world than did Tommy Farr against Joe Louis at Yankee Stadium here . . . Farr's gallantry was complete, and a scathing answer to the American critics who, almost without exception, held him to be no sort of fighter, certainly no foeman worthy of the negro's steel.'

Schmeling, at the height of his arrogance, suggested, 'Shirley Temple has as good a chance' – although this was as much a snide shot at Louis as it was meant to deride Farr.

Bennison, nevertheless, felt justified and comfortable in declaring to his readers back home, 'I say without hesitation that Farr proved himself the better, cleverer and more resourceful boxer.'

Farr finished with two cut eyes, Joe with two bruised hands. The champ came down the stretch strongly after an indifferent middle patch and the judges were impressed, albeit by wildly varying margins. Art Donovan, the referee, gave Louis thirteen rounds, Farr just one, with one even; the other two judges saw it 8–5–2 and 9–6 for Joe.

Farr's manager, Ted Broadribb, did not complain about the decision. Neither did the Tonypandy Terror himself. 'Are you satisfied that I have not let either myself or my country down?' is how he humbly responded when Bennison put it to him he'd come damn close to becoming the first British world heavyweight champion since Bob Fitzsimmons.

And even Bennison had to concede, with all the reluctance of an expert whose prediction hadn't gone exactly as foreseen, 'There was, according to my reckoning, only a fractional difference in favour of Louis at the end, and it says much for the sportsmanship of Farr that, when he was declared the loser, he took the verdict without the least quibble.'

Joe's purse for seeing off Tommy was $102,578, a little over a grand less than he'd earned for beating Braddock.

After Farr, they set 'em up for Joe, and Joe knocked 'em down. It looked like great business. For Joe. For Jim and his Joe. And for Uncle Mike.

However, Joe Gould sniffed dismissively at the champion's next purse: $40,522 for a cakewalk against Natie Mann in the Garden in February of '38. He was similarly unimpressed in April, when Joe spent a mere five rounds getting Harry Thomas out of the way for a paltry $16,659.

Schmeling, meanwhile, waited and fumed – and turned the Atlantic into his personal highway as he criss-crossed to force a showdown with the man he'd beaten in '36, the acknowledged world champion. There would be no complaints about the purse this time.

It was a marketing man's dream. Joe was fighting for every good guy who ever lived. His autobiography, published in 1947, was a whitewash typical of the genre of light-hearted and wholesome accounts at the time. The final chapter, entitled 'P.S. – WHAT AMERICA MEANS TO ME', reflects the enormity of the task assigned to him as a mere fighter.

He recalls meeting President Roosevelt at the White House just before the rematch.

'You know, Joe,' the president said to him, 'America is never supposed to lose.'

'I know, Mr President,' Joe said. 'And I'll take care of that this time!'

The media frenzy, from left, right, German, black and Jew, was unrelenting. After so many postponements, so much hassle, so many tactical manoeuvrings and double-crosses, it had to go ahead. There was no avoiding the German now.

And so Louis–Schmeling II took place in New York on 22 June 1938. Joe's purse was $349,228, a remarkable amount of money – for a remarkable fight.

The 124 seconds it lasted were burdened with greater poignancy than anyone then or since has attached to a mere boxing match. And the 'wider significance' of the occasion was not lost on the 70,043 fans who paid to get into Yankee Stadium that humid midsummer's night, nor on the millions who listened to English, German, Spanish and Portuguese commentaries on radios around the world. If it were possible to recreate an event of vaguely similar global consequence today, a fight between two men representing good and evil in such simplistic, cartoon terms and in circumstances of such heightened international tension, there surely would be billions entranced all at once.

Schmeling had the best view of some of the punches, but not the ones that mattered – particularly the one to the kidneys he later whinged was a foul, the one that Joe sank into his pale, untended trunk just below the ribcage in the first round and from which Max could not recover. In reality, it was a bolt-like right to his chin in the early seconds that did the damage; thereafter, the blows struck all parts. There was nothing illegal – in New York, at least – about the kidney killer that took away Schmeling's resolve, though.

Max went down. Art Donovan, again the referee, applied the count. Max got up. The terrible but beautiful assault continued. Joe was cold, balanced and merciless. He just picked targets and let his gloves go. The hesitancy he'd shown against Farr had gone. A winded and bewildered Schmeling could not get out of the way, even when he turned sideways along the ropes like a boy being bullied. For Americans watching and listening, for others with an interest beyond the boxing ring also, this was retribution of the sweetest kind. The bad guy was getting his licks, good and proper. Dazed to the point of incomprehension, Max wandered like a lost sheep back into the storm and his legs had not a drop of strength in them to keep him upright as Joe slayed him like a righteous knight. The German swayed, tottered and sailed south as Joe's fists rattled jawbone, brain and spirit simultaneously. Max, clinging to the edge of the battleground, was up at five, but in his own hell. The white towel floated in. Donovan ignored it, in accordance with local statute, and then applied his own mercy.

It was a brief proto-war, shown so many times since as to be fixed irremovably on the brain like a birthmark, a round that made America feel good about itself again, a round that put the world to rights, it was claimed. After the indignations and hardship of the Depression, symbolism hung heavy in every punch Joe threw.

When it was over Max could look back on exactly two punches of his own. He was spent as a heavyweight force. Joe, who'd thrown and landed maybe a hundred, was reborn. And so the Joe Louis story could resume.

Schmeling had the manners and judgement to observe years later in his autobiography, *Memories*, 'Every defeat has its good side. A victory over Joe Louis would perhaps have made me into the toast of the Third Reich.'

Putting the best gloss on it, maybe Schmeling learned

something from Louis that night. Maybe he learned that with a decent and honest hiding sometimes comes humility and respect. Nevertheless, however tempting it is to paint the fight and the result as a blow against Hitler and a triumph for democracy and the American way, the contest had less to do with the rights and wrongs of their respective ideologies (if indeed they would even have called them that) than with two fighting men testing themselves to the limit for a considerable amount of money.

Joe, simply, was too good for Max. Defeat did embarrass the Nazis, of course. And how good was that? But it had no tangible effect on their evil intent. It did not stop them annexing the Sudetenland in October. It did not delay their invasion of Poland the following September to start a world war. It hardly shamed them into treating Jews, Gypsies and communists as human beings. But it did give Max Schmeling the opportunity to acknowledge that Joe Louis was the best heavyweight in the world and, by so doing, secure his own place in history.

It did also, briefly and dramatically, make America and what would come to be called the Free World feel uplifted. But soon enough the clamour faded, and everyone got back to business. It was, in the end, just a fight. As Mike Jacobs knew, it was business.

For Joe, and others, the money kept rolling in as the world slipped into its second global conflagration: John Henry Lewis, 25 January 1939, at Madison Square Garden, earned Joe $34,413; Jack Roper, 17 April 1939, at Wrigley Field, $34,850; Two-Ton Tony Galento, 28 June 1939, at Yankee Stadium, $114,332.17 . . .

There is confusion to this day about the exact terms of the deal Gould did with Jacobs for Braddock. Some say it was 10 per cent of the gate whenever the title was contested at the Garden. Another theory has it that Braddock got a tenth of the overall promotion, no matter where it was held, as long as Jacobs and the Twentieth Century Sporting Club were involved. My guess is Uncle Mike dipped into the champion's purse to meet his part of the deal, writing it off as expenses, one of the fight game's oldest dodges. Towards the end, Jacobs, his health failing, tired of the arrangement and their lawyers swapped expensive letters.

As Lucky Jim remembered it years later in conversation with the writer Peter Heller: 'I got 10 per cent of the promotions involving any championship because once Louis won the fight, Mike Jacobs, who controlled Louis, controlled the heavyweight division, and he had control of that title. But, if Louis got knocked out, we didn't make it with Louis, we made it with the promotions the next 10 years, regardless of who was champion. As long as Jacobs promoted that fight, we were in for 10 per cent, like an annuity. We might have got one hundred fifty thousand out of it over the 10 years. Which wasn't a bad annuity.'

6

DISTANT DRUMS

For Europe and their allies, the real fighting, the irreversible descent into global conflict, began at 4.40 a.m. on 1 September 1939. Schmeling was not among the Luftwaffe airmen who hit predetermined targets in Krakow, Łodź and Warsaw that dawn. But he, and the rest of Germany, knew that a terrible beast had been let loose. Five minutes after the planes took off, the German Navy was bombing the free port of Danzig. By 8 a.m., the Wehrmacht had moved on the village of Mokra, only to be repulsed, a rare Polish victory in what was to become a nightmare occupation for the next six years.

While Max's ring cachet had been seriously diminished, he nevertheless was retained as a faded German hero from the last fragile days of peacetime. Not even that peripheral clout, however, could save him from the inconvenience of being drafted and serving as a parachutist in the invasion of Crete in 1941. Later, it is said, he gave shelter to Jewish refugees. By then, he had been parked in the relative safety of a military hospital in Ulm. His war was a decent enough one.

Joe, meanwhile, boxed on: Bob Pastor, 20 September 1939, at Briggs Stadium, Detroit, $118,400; Arturo Godoy, 9 February 1940, at Madison Square Garden, $23,620.21; Johnny Paycheck, 29 March 1940, at Madison Square Garden, $19,908; Godoy II,

20 June 1940, at Yankee Stadium, $55,989.04; Al McCoy, 16 December 1940, at Boston Garden, $17,938; Red Burman, 31 January 1941, at Madison Square Garden, $21,023.16; Gus Dorazio, 17 February 1941, at the Convention Hall, Philadelphia, $18,730.70; Abe Simon, 21 March 1941, at the Olympia Stadium, Detroit, $19,400; Tony Musto, 8 April 1941, at The Arena, St Louis, $17,468; Buddy Baer, 23 May 1941, at the Griffith Stadium, Washington DC, $36,866; Billy Conn, 18 June 1941, at the Polo Grounds, New York, $153,905 . . .

They called it the Bum of the Month Club. Which, Conn and Godoy aside, was not far from the truth. It was like a butcher's shopping list, with some cuts fresher than others. And, as the champ stood over each of his slain quarry, Joe Gould could hardly contain his happiness.

In America, the sense of removal from the Old World, embedded in the national psyche since the earliest days of colonisation, seriously delayed for the second time in the century their entry into a world war. The cries of the Anti-Nazi League who'd demonstrated against Schmeling being allowed to represent the Third Reich against Louis in New York in 1938 had little impact now in the White House. The president had urged Louis to stand up and fight for his country against a German in the ring then, but Roosevelt was not able to persuade his countrymen so easily that taking up arms for old allies was worth it now. It took the Japanese bombing of Pearl Harbor in December 1941 to change that. The bombs fell on American ships just six months after Joe came from behind to knock out Billy Conn in the thirteenth round and keep his title.

A month after the day of infamy, as Roosevelt memorably called it, he gave the thoroughly outclassed Lou Nova an awful beating over six rounds at the Polo Grounds in New York, earning another $199,500. In January of '42 at the Garden, he went to work on Max Baer's big brother, Buddy, for $65,000, a rerun of an earlier farce; in March, he gave Simon a second walloping at the Garden, this time for $45,882.

Later that year, Joe joined up. He was spared the chore of killing or dying for democracy, though, and sold war bonds for his country instead, rousing the troops in his stumbling, inarticulate but sincere way. For the duration, he had one allegedly serious contest, against a guy called Johnny Davis in Buffalo over four rounds. Davis, knocked cold after just fifty-three seconds, could nonetheless tell his grandchildren he once fought Joe Louis for the world heavyweight title because, according to the rules of the New York State Athletic Commission, the champ's title was on the line every time he stepped into a ring. How Davis, who was knocked out nineteen times and lost twenty-one of his twenty-six fights between 1944 and 1946, ever got to share canvas space with one of the finest fighters of all time remains a mystery. And it did Joe's credibility no good at all after his string of one-sided defences. It did not seem to bother him, though. For Joe, it was business, an interruption to his new obsession, golf.

The rest of World War II, Joe travelled the country, went to Europe, boxed exhibitions, had his face on billboards, he was on the screen and on the radio. From fights for the Army and Navy Emergency Relief Fund, he raised $100,000 – all of it taxed. It was a debt that would one day crush not only his finances but his spirit.

Had he been asked, Joe would have killed Germans for a buck private's pay of $1.25 a day. That is pretty clear from what he said at the time, and later. But Joe knew he would not be asked to go into action, even alongside fighting men who adored him. America did not do that to their celebrities. Also, he was black. There was a residual prejudice that blighted the American war effort, as black soldiers, sailors and airmen struggled with the dilemma of defending a country where they were still considered second-class citizens. It was as if blacks were so far down the social and cultural scale, they were barely worth sacrificing on the battlefield.

Into that very argument stepped a man who would later play a pivotal role in Joe's life, and that of the Mob in boxing.

Truman K. Gibson Jr, to give him the handle he always insisted on, was born in Atlanta the year the *Titanic* went down, 1912, two years before Joe was born. He grew up in an educated and proud black family in Columbus, Ohio. His parents were articulate teachers prominent in the black community – and not fazed by celebrity of any kind, black or white. His mother, Alberta, had some memorable personal run-ins at their house with W.E.B. Du Bois. He was the renowned black intellectual and commentator who'd stepped into the debate over Jack Johnson's defeat of James J. Jeffries back in 1915. He was also the first black man to earn a doctorate from Harvard (he would ease Truman's father's path into that institution) and a man whose ego was hardly prepared for Alberta's sharp tongue. When he came calling one day and complained he'd only asked for 'half a cup of tea', Alberta handed him back the full cup and reminded him she was a busy woman. 'It saves me a trip back to the teapot,' she said. 'Take it or leave it.'

His father's grandmother, whose full name Truman never knew, was a part black, part Seminole Indian – and, like Joe's grandparents, a slave. So young Truman needed no history lessons in the oppression of blacks, or how to strive to rise above that oppression. His father, an independent and strident thinker, left academe to become a leading figure in the insurance business, in Atlanta and later Columbus. If you were looking for a pattern, there might be one in how the son, like the father, drifted from advocacy and intellectualism towards the convenience of pragmatic moneymaking.

Truman attended integrated schools in Columbus but recalls in his autobiography, 'We were distant from the other kids; I had no white friends at school.' He played on the school football team but he and the only other black player dined separately. There was even a black YMCA in town. Segregation thrived in all corners of the city named after the white man credited with discovering the country. So he moved to Chicago. It was a move that would have consequences beyond his immediate career.

Gibson had also done his bit during World War II. He was as much of a ground-breaker for his race as was Jack Johnson, although in a more subtle way. He had an ego – for example, the use of the full, drawn-out moniker of Truman K. Gibson Jr – but he had reason to be proud of his achievements. For all his later weaknesses, when he chose to do business with Jim Norris and his cohorts attached to the International Boxing Club, Gibson could go to his grave content with his contribution in tearing down Jim Crow prejudices in the armed forces before and during World War II, a time when bigotry was far more entrenched than now.

He was a big barrel of contradictions, noble and weak, intelligent and – conveniently, perhaps – naive. He made the best of whatever situation he was in, but failed to see that he could have avoided being in some tight corners in the first place had he not let circumstances drift. Truman Gibson might have been boxing's ultimate pragmatist.

In the autumn of 1940, Gibson, one of his country's few prominent black lawyers, was summoned to Washington to act as an advocate for African American soldiers. For five years he served as an assistant to Bill Hastie, the civilian aide to the Secretary of War. He saw prejudice against black soldiers everywhere 'abused, assaulted and even murdered by white civilians in the south', as he recalls in his autobiography.

When he and his wife arrived in the capital from Chicago, they were angered to learn segregation was deeply embedded in the Washington white mindset, from top to bottom. Even shop staff refused to let Mrs Gibson try on a pair of shoes. But Gibson's time had come; there were stirrings of black awareness on Capitol Hill as President Roosevelt, repairing the damage left by Hoover, was gearing up for his third term and leaning on the black and liberal vote.

Gibson marshalled black opinion-makers in the media but felt powerless sometimes in trying to shift perceptions in society at large. As America stirred itself to fight fascism, at home its citizens were prepared to countenance all manner of segregation, from shoe shops in Washington to schools and buses in Alabama, and all the way through the military, where Negro-only units were still in place.

Gibson wanted to change attitudes but came up against brick walls everywhere. 'The army is not a sociological laboratory,' the chief of staff, General George C. Marshall, told Gibson when he pointed out the many anomalies he discovered. But Truman persisted. He listened and learned. And argued. He walked out on intransigent opponents to his initiatives. Gently, he twisted arms. He even succeeded in getting blacks into the air force, long considered a bastion of whiteness and privilege. It was a small victory, but a significant one.

Gibson kept chipping away at the institutionalised racism in the military, like a pesky flyweight jabbing, jabbing, jabbing. His integrity and demeanour won him friends in the right places and, slowly, views moved in the direction of fairness and equality. It was not a tectonic shift, but its subtleties would be felt for longer than any single major eruption. Gibson demonstrated then in the halls of power the mental agility he was to bring to bear in that other bear pit, professional boxing.

It was a curious battle to fight: striving to give his black brothers and sisters the right to go and be blown up for the country that denied them so many basic rights and freedoms. He received one letter from a black soldier at Camp Lee in Virginia that summarised the dilemma: 'The prisoner of war gets much better treatment than we do, even when they go to the dispensary or hospital, and it is really a bearing down to our morale as we are supposed to be fighting for democracy. Yet we are treated worse than our enemies are . . . If something isn't done quick, I am afraid a great disaster will surely come.'

Fame in sport did not spare black servicemen from prejudice. Jackie Robinson, who, after the war, would become the first black player in Major League baseball, did his basic training at Fort Riley in Kansas. Towards the end of his time there, he overheard a white officer call a black soldier 'a stupid black nigger sonofabitch'. Robinson intervened, thinking his standing as a rising baseball star might carry some influence. 'That goes for you too, nigger,' was the curt reaction. Robinson threw an angry right hand and knocked several of the officer's teeth out. Luckily for Robinson, Joe Louis, who had met Gibson in 1935, was by now also at Fort Riley. He informed his lawyer friend of the incident.

As Gibson put Robinson's defence to the officer in charge of the investigation, Louis intervened in a way that those who knew him might have imagined was beyond his simple ways. 'General, you have to do a lot of entertaining and I took the liberty of delivering a case of wine to your quarters,' the heavyweight champion of the world told him. 'This is not any bribe or whatever, but I would like for my friend, Jackie Robinson, to finish his course.'

Robinson graduated from the officer candidate school shortly afterwards. However, he struck more bigotry soon enough. On boarding a bus at another army camp in Texas, he was told, 'Nigger, get to the back.' The combustible Robinson refused – and grabbed the driver's quickly drawn pistol, raking his mouth with it. Louis wanted Gibson to intervene again, but it wasn't necessary. Robinson had a result of sorts; he was honourably discharged.

These were the times in which Gibson flourished as an advocate for his people. There were hundreds of similar cases. Frank Capra had been co-opted to make a series of propaganda films for the War Department under the umbrella title of *Why We Fight*. As part of his remit, Gibson enlightened the famous director about what was happening to black units and Capra subsequently made *The Negro Soldier*. *Time* magazine wrote that it was 'just about the first time in screen history their race was presented with honest respect'.

Throughout the war, Gibson continued his fight against racism. He had done a fine job for his people. And he always looked out for Joe Louis, not because he was the heavyweight champion of the world, but because he was a black man well placed to inspire others not so advantaged.

When they first met, in Gibson's Chicago office a year before Joe became world champion, Joe was with his manager, Julian Black, and Black's attorney, Irvin Mollinson, who knew Gibson. They asked Gibson to entertain the fighter while they discussed some property deals, and the two became firm friends instantly.

The meeting was a collision of circumstances. Such was the size and contracted influence at the high end of the black community it was perhaps inevitable they would meet one day anyway. Gibson did not see it as a business opportunity for himself, as he explained: 'I, like every other African American, was intrigued by Joe and the possibilities he represented. I had never seen a boxing match, had no interest in the sport, and of course had no way of knowing the huge role it was to play in my life. Still, I knew chance had placed me in the presence of someone special who might make history.'

Gibson, obviously, knew his history. He knew Joe had won the Golden Gloves, turned pro and won twelve quick fights. He was no boxing fan, but he could hardly have missed the buzz Joe was creating across America with his fists. And he knew that, where Johnson had angered the white establishment with his arrogance, Louis was being coaxed into a more compliant role by Black (who'd been a business partner of Johnson's), advised to walk quietly, make no fuss. In short, to be a good American. It was, according to the view of Gibson and many black leaders of the time, a great opportunity to knock down some barriers. 'Now Joe had his foot in the door,' Gibson writes in his autobiography, 'and under the right circumstances it appeared just possible that he might achieve the breakthrough. Black America sensed that something extraordinary was about to happen.'

And, of course, it did – thanks to the singular genius of Joe Louis. He was, at the beginning, a shining hope. However, nearly everything else that happened to the fighter in the years afterwards represented a criminal waste of his potential and his contribution.

When the champ got out of the army in 1945, he remembered the quiet guy from Chicago who'd advised him how to get Jackie Robinson out of a hole while they were in army camp in Kansas. So he asked Gibson to manage his boxing career. Gibson at first declined, recommending he return to Julian Black. But Joe fell out with Black when he failed to facilitate a sizable payment to his estranged wife, Marva, and it all got a bit messy. Gibson, somehow, got caught up in the acrimony when Joe, who often became confused and paranoid about the complications in his life, accused him of taking Black's side, which wasn't so.

They made up eventually and Gibson worked comfortably enough as an attorney for Louis with Joe's new manager, Marshall Miles, a well-connected boxing face, who took over Joe's career from his friend John Roxborough, when Roxy went to prison. That's how a new partnership was forged in the latter stages of Joe's career. It was not entirely a bundle of laughs. And, in the end, Gibson would have much to answer for.

When I met Truman Gibson in his Chicago offices in 2004, he was the most amiable and erudite company, a venerable survivor and a rare first-hand link to the Joe Louis era and everything that that entailed. He talked candidly and knowingly about all aspects of the boxing business, from how he met Louis to what he knew about Sonny Liston. Yet he remained on guard.

Nick Tosches had not long released what I thought was an excellent book on Liston, *Night Train*, but Gibson wasn't wholly convinced. 'Not a bad read,' he said of Tosches's exploration of Liston's Mob affiliations, 'but he didn't get it quite right on everything.' This was the knowing remark of someone who could obviously keep his counsel.

I have to say, Gibson looked a contented man in his storefront office. These were his people, even if most of them didn't have the advantages that had blessed his youth. He was their hero. Like Joe had been. Right up until his death at the age of ninety-three, Truman was taking calls from clients. His wife had died a year earlier, but he was determined to keep doing the job he felt was his destiny. His CV was impressive. He served on two presidential advisory committees and was the first black American to win the Presidential Medal of Merit. He was, by any standards, a remarkable if imperfect man.

In his autobiography – co-written with Steve Huntley, the editor of the *Chicago Sun-Times* op-ed pages – Gibson, not surprisingly, gives us an account of his time in boxing that suggests he saw or heard very little he would characterise as untoward.

'Truman Gibson, boxing promoter. That's a line I never dreamed I'd find myself in,' he says.

Ultimately, it is hard to know what to make of him, though. As Bert Sugar, king of the angled hat, unlit cigar and a boxing sage of enduring amusement, says, 'He was a sweet, sweet man. He did as best he could for Joe.'

Maybe he did. And maybe his best wasn't good enough.

Joe Louis smiled quietly on the August day America went crazy celebrating the end of yet another world war. It was a victory for the forces of good, for liberty and democracy, the first just war, most agreed, and one that united people around the world against the spectre of subjugation. The insanity of Hitlerism galvanised the higher instincts of humanity. The world, for the second time in the space of twenty-seven years, had been freed again.

Freedom had not always been Joe's for the taking. Nor was it now, totally. What little he'd got, he'd hit hard for. Not against the Germans, but in his own country, the Land of the Free. Down his own streets, across the rings of America, with dignity and without fuss. It did not come easy. Joe, unlike the rebel of his people, Jack Johnson, was a hero who had to smile at the right times. It was not in his nature (nor did it suit those around him) for Joe to complain publicly about injustices or petty slights. He had grievances. Many of them. But he settled them quietly. He would never tell the world what Joe Gould did to him. Nor would he speak about Carbo, Palermo, Norris or his friend and legal confidant, Truman Gibson. Joe would play the part of brave soldier, the ultimate fighting man, a good American. Right or wrong, Joe Louis did not see it as his assignment in life to preach. Besides, he would not have been very good at it.

The obligation of athletes, especially fighters, to prove their physical courage in time of war was an enduring one, stretching back to World War I when Jack Dempsey, who was rising towards the championship, came under pressure to sign up. He didn't, but claimed he'd tried. Doc Kearns, always scheming on his and Jack's behalf, even dummied up a picture of Dempsey working in a shipyard as part of the war effort. Only trouble was, the champ was wearing street shoes. In World War II, Jack would try to make up for this lack by joining the Coast Guard. It is said he offered to fight at Okinawa in 1945. He was forty-nine at the time.

Joe never shot a bullet, either, at a German or a Jap, but he was there for the boys when asked, gauche but willing.

Joe Louis Barrow, as he was born back in 1914 in a shack in Alabama, just two years after his father Monroe had been admitted to a mental hospital, was the original GI Joe, and America adored him. They would not know until the damage had been done how much he'd given for their entertainment. And they could not guess there was even more to come.

7

THIS ISLE OF JOY

When all the GI Joes came marching home again, it was to New York, a city that was made for parades. No metropolis has been so perfectly shaped to accommodate the celebration of man's triumphs, a canyoned, elongated thread of bumpy tarmac, a giant concrete cake of fun. Be it for the birth of an Irish saint, the return of another crazy aviator or the end of a global conflict such as World War II, the ticker tape always fell well on Manhattan. And never did it fall more beautifully than on that Wednesday of 15 August 1945, VJ Day. Americans might have entered the war reluctantly, but they punched their weight, with their lives and their dollars. Everyone knew that without them the cause was probably lost. It was a time, long forgotten, of sacrifice freely given. Evil was clearly defined. Good was easily understood. It was a war, more powerfully than any other, about justice.

There were other, if lesser, celebrations to be had that day in New York City. While the fighting men in uniform were relieved to be home and getting ready for their big parade, Billy Graham woke up in his apartment in lower Manhattan and stepped out to get the papers. Sure, he was going to watch the guys march down Broadway. He was as patriotic as the next guy. But he was keen also to see what the fight writers said about his own fighting two nights before out at the Queensboro Arena in Long Island City.

Billy was on the way up. People were starting to talk about him. And not just in his Irish neighbourhood.

The show wasn't a big one but his eight-rounder against Johnny Rinaldi was top of the bill. He'd been promised a big fight in the Garden. 'Any day now, Billy,' Mike Jacobs kept telling him. 'Don't worry about it. You'll get your shot. You're a good kid. Be patient. We'll look after ya.' Graham was unbeaten in fifty-six fights stretching back four years; he could not afford to slip up now. He didn't that night at the Queensboro, either. How could he? Rinaldi, from the Bronx, was washed up. He had been around forever. He'd had seventy-eight fights and won only thirty times. People had been beating Johnny up almost since the day he started fighting, back in 1936. He could punch and he'd had his moments, mainly against other tomato cans. He'd even had a few of his own nights in the Garden, down the card, and they'd give him the call again before he wound up his career. Mostly he was put in there to lose. Not many lost as convincingly as Johnny Rinaldi.

Billy was a live item, though. He stopped Johnny in the fourth and the papers were kind. He'd had five Garden fights himself. All prelims. All wins. Still, whatever the nice things the fight writers said about him, Billy was getting nowhere fast. His was a boxing life of four-rounders, six-rounders, promises. 'You'll get your shot, kid.' Billy would go to sleep with those words ringing in his ears. He wanted a big one, a ten-rounder, a pay day, a crack at the welterweight title. He knew what you had to do to get all that; you had to know the right people. And you had to do what they told you. Rocky Graziano, he grew up not far from Billy, and he knew those guys.

In the bright August of his life, the fight reports tucked in the right-hand pocket of his worn tweed jacket, Billy had a moment to forget the fight game. 'Let's get down and see what the hell's going on, Rock,' he said. 'Those damn GIs and sailors, they're probably stealin' our broads . . .'

New York was in orgasm on that sunny Wednesday, through the innocent virgin cheers on up to Times Square at the junction of Seventh Avenue. This was Victory over Japan Day. This was New York. This was the right minute of the right hour to be standing in the middle of the city that was at the heart of Americanism. Billy and Rocky pushed their way into the throng. 'Hey, Billy!' he heard, as he squeezed into Greeley Square, soldiers, sailors and fly boys moving along the street beside him. 'You showed that bum Rinaldi!'

In narrow streets all around, hubbub, hope and relief filled the bars as the city embraced its war heroes like the ancient Romans loved their fighting champions.

FDR was dead. The shrewd and homely president who wrapped himself in a car blanket as he guided the fortunes of the Western world from his wheelchair, the New Dealer and internationalist who'd taken them into the war, allegedly against the will of his fellow Americans, didn't live to see the end of World War II. Harry S. Truman was in charge now. He was the president who dropped the Bomb. He would give life to the Cold War. But, first, before America even had time to reflect on the carnage that had unfolded over the previous six years, before they counted the cost in lives and dollars, there was drink to be had, dances to jive, songs to sing.

A belief in the future rippled through America that was missing in nearly every other country in the world. In Britain and Europe, in the old colonies and in the undeveloped world, even in the controlled Eastern bloc, post-war weariness hung on like a flu virus. They had seen off evil, but the good times were a slow train coming.

In America it was different. The nation might not have been in the rudest of financial health, but the engine was idling promisingly. New gadgets, fresh ideas and relief at returning to their cosseted normality were underpinned by a spirit of adventure, all of which inspired optimism. Americans did not have Old World baggage. Germany had failed to launch their Third Reich; Americans, meanwhile, had quietly confirmed they were the new Romans.

The United States had not so much survived the war as used it to resuscitate the world's most vibrant economy, one that only a decade earlier had been wheezing like a punch-drunk pug in a four-rounder. It was fitting that America should experience the extremes. America was the toughest, the strongest – it could take the heaviest blows. When the New World sent shiploads of cash to the Old World to prop up the Marshall Plan in the fifties, it was no more than they could afford. It was a gargantuan bribe not just to rebuild a shattered continent (and provide American companies with some lucrative contracts) but to construct an invisible wall and shut out the perceived evil of communism. America was getting into its stride as a world power like never before. A new empire was growing before the world's tired and envious eyes, on cinema screens, the radio, in books and magazines, and on 78rpm discs of crackable vinyl. If you could get them.

America was the fulcrum of modernity. What happened in America would, from now on, set the tone for the rest of what was regarded as the civilised West. Indelible lines of morality, philosophy, style and materialism were being drawn in those dramatic post-war months and the rest of the late forties, after the Axis was dismantled.

In New York City on the day the war ended, old Harlem bellowed the blues, brass bands gouged the summer day downtown and, through the night in caverns in the Village, bebop hipsters clicked and nodded to the melodic sax of their godfather Coleman Hawkins, and their new god, Charlie Parker. Rock 'n' roll, whether anyone knew it or not, was on the way.

There was also an anthem to match the mood, written twenty

years before by Richard Rodgers and Lorenz Hart, immortalised by Ella Fitzgerald. This was the place she sang about. This was Manhattan, this isle of joy.

Music had started to swing during the war, thanks to Glenn Miller and other big bands, and the dance floor rocked, as boys and girls gaily lost themselves in free-form, care-free happiness. When the war was over, they turned up the volume a little, and got down to some slick lyrics, about love, youth, hope and fun. This was to be the age soon of the teenager, a phenomenon that would drain parents' pockets and frighten old folks. Billy knew what they were writing about. Crisp-skirted girls with perfect teeth swished their shiny ponytails in the faces of youths who had seen death thousands of miles away and wanted nothing more than a girl-shaped bottle of Coke and a bit of boogie back home.

All down Broadway, Billy grinned like the kid he was. Maybe Uncle Mike was right. He'd get his shot. Just like Rocky. Everything would be fine . . .

When the parade was over, the great chroniclers went to work on what had been a day nobody would forget in a hurry. The images were seared in the memories of those, like Billy, who'd cheered and drunk and laughed and believed in some sort of better future. For those who didn't make it to Times Square (in the era before TV), there were artists and reporters in good numbers to inform them about a magic day.

Lingering high in the brain above other memories, there is 'the photo'. It froze a skinny sailor bending back a nurse of possibly recent acquaintance in a U-shape and landing on her lips the grateful kiss of his and many other ships. This was breathless catharsis, brought to the world on the pages of *Life* magazine through the keenly tuned lens of Alfred Eisenstaedt. The great photographer said in 1995, when he was a mere ninety-six, 'People tell me, "Oh, you have taken pictures of Loren and Monroe!" But this is what they know me for.'

Such was the impact of a single image. This was the definitive representation of VJ Day, in an age when recollections were put down simply and definitively, in black and white mostly, and never questioned. This one was cherished between the covers of magazines that faded in the light but not in the retelling. Like good Shakespeare, it was a snapshot of its time, timeless and incorruptible.

The picture did not just capture a moment, it reflected a mood. It felt right. It was a picture of love after six years of hate. It was something ordinary that anyone could understand, yet it was aspirational – and, naturally, in a land of make-believe, people could understand that too. Because it might have been a grab from a Gene Kelly movie, so harmonious were the melding thighs. All seemed to click in sync with Alfred's old camera.

These homecoming champions were Adonises with lean muscle, heads brimming with dreams, chased and loved by equally lovely, equally hot young women eager to crank up the fires of the good old US of A again. And, like generations before them, they wanted to shape the world in their image.

They were tired of dancing with death – and Eisenstaedt's famous photograph seethed with sexual potential, yet it was not lascivious. After the horrors they saw in Europe and the Pacific, young American soldiers were coming home to what they hoped would be some version of the innocent world they'd left behind. They wanted fixed certainties.

Nine days later, on Friday 24 August, they were still cleaning up Times Square. On Eighth Avenue, it was fight night at the Garden, a sure sign New York was more or less back to normal. Billy was there, but not fighting. He got a ticket in the cheap seats, to see Rocky in the main event, against a guy from the Mob town of Elizabeth, New Jersey, Freddie 'Red' Cochrane. Graziano was on his way already, twenty-three, only a year older than Billy, and already filling the columns of the papers, getting the calls from boxing's movers. Rocky had been a pro four years and his fight log was 38–6–5, his five defeats all in the Garden, out of nine starts there – but the fans loved him. Rocky gave them a show every time he got in the ring.

He was as up for this fight as he was for any. He had been told he was not far away from a title shot. Rocky got nervous before a fight, like Billy, but he knew how to use that energy. He saw Billy heading up the stairs to the auditorium and called him aside.

'Hey, Billy.'

'Rocky. Good luck tonight. You've got to get a shot at the title soon.'

'That's what they keep tellin' me, Billy. Say, you heard about this Cuban, Kid Gavilan?'

'A little. Won thirteen straight, they say. Can bang a bit, too. One to avoid, Rocky. At least until you get to make some money together, eh?'

'Right, Billy. Gotta get Red outta the way first. Talk later.'

Gavilan was on the lips of a few fight-game wise guys in '45. The *Ring* had been keeping an eye on him from his very first fight, back in 1943, when he won well against another kid making his debut. 'Gerardo Gonzalez (later Kid Gavilan), 122lbs, a sweet little boxer, outpointed Antonio Diaz, 122lbs (Cuban records name his opponent as Baby Chango) in six rounds and won the plaudits of the crowd,' is how they reported this faraway fight – for which Gavilan earned $12.

Gavilan's story is classic. He was born Gerardo Gonzalez in Palo Seco, a village near the cane fields of Camaguey, in 1926, and worked in a kitchen from the age of ten. Gerry, as his friends called him, was so small when he started boxing for the local club he was assigned to the *guasasas*, or 'little pests'. But he won in the country and headed for the capital, to the annoyance of his mother.

In Havana, he struggled to find anyone to take him on until he met a small-time trainer called Manolo Fernandez and an ambitious boxing enthusiast, Fernando Balido, whose main employment came from his café on the corner of Animas and Escobar. It was called El Gavilan, in Spanish 'the hawk'. And that was to be the Kid's name for the rest of his life.

He boxed around the two main venues of Havana, the Arena Cristal and Palacio de Deportes, over the next two years and, in July 1945, knocked out the experienced Joe Pedroza to win the Cuban lightweight title.

The word in August was 'The Keed' – who had filled out somewhat – was on his way to New York. Looking for Rocky, and any other decent welterweight. He'd never fight Graziano. But he would eventually fight Billy – after falling in with the Mob. It was almost ordained, given his Latin connections.

Graziano knew the Mob through his Italian roots. That is the way things worked then. In his early days, he seemed comfortable enough with the attention of these 'businessmen'. He was a fighter, like La Motta, Graham and Gavilan. What could he do about it?

The Rock was some item, a tough but sensitive man, abused as a child by his alcoholic father who boxed as 'Fighting Nick Bob' Barbella. Born in Brooklyn, Rocky grew up in Little Italy on the Lower East Side. As an amateur, he learned to box at Cus D'Amato's Gramercy gym, on the corner of 14th and Irving Place; however, the wise guys got in Rocky's ear when he still didn't know which way was up and he left Cus, who was not playing ball with the boys at the time. D'Amato, and a few others, reckoned Rocky could make it to the top as a pro. Now the Mob had stolen the kid away from him. Cus ever after trusted nobody in the fight game and, it is said, slept with a gun under his pillow, so suspicious was he of the guys who cruised New York in their shiny black cars. Whether this caution was merited, as his protégé José Torres maintains, or has become a myth invented by D'Amato to enhance his reputation, is as unclear as so many of these beguiling stories in the boxing business.

Certainly Cus was what Americans like to venerate as a self-improver. He only made it to the eighth grade but, at nights, he would roam the shelves of the New York Public Library on 42nd Street, reading anything that caught his eye, but mainly books on history and psychology. He knew how to throw a sentence together.

On the face of it, meanwhile, this unusual and mysterious man, with the white hair and the bowler hat, the fast tongue and the convoluted philosophy, was an inflexible enemy of the Mob, from as long ago as the late forties. His friends say this sprang from his righteousness; others reckon he was just plain stubborn, that he refused to cooperate with them because it would compromise his own sense of self-worth. Still others reckoned he just never got a good cut on the deal from the wise guys. (Many years later, when Cus discovered another world title prospect, the shy, brittle and often brilliant heavyweight Floyd Patterson, he was to learn just how tough it was defying the will of the gangsters, whose own main man, Sonny Liston, moved darkly through the heavyweight division like a mugger after midnight. So Cus made sure Floyd made all his big money before Sonny got to him. How right he was.)

Whatever the later speculation about him, to D'Amato, the Gramercy in the forties was more than a gym; it was a shelter from the bad guys, from all that was not beautiful in life. He'd learned the hard way that the streets were not all romance and risks. As a kid he got in a street fight and lost his left eye. The injury stopped Cus from following his older brother Jerry as a pro fighter. Jerry's boxing didn't get far, anyway; when he was seventeen, the local cops gunned him down and left him dying in the gutter of the Bronx streets of notorious Classon Point. So young Cus became a teacher, not just of boxing but of life. He was equipped to do so through experience, although he said much that carried the wisdom of an intellectual. The one instinct he held dearest was respect – respect for other people's talents as well as their ability to expose your own frailties. He trusted nobody, except maybe himself. It was some achievement for him to retain any human warmth at all, but he did.

As Kevin Rooney, one of his many one-time rascally pupils recalled many years later, 'Back in the forties, Cus kept the Gramercy open 364 days a year. On the 365th day, he'd give a Christmas party for all the kids in the gym and would buy them presents. There'd be food and drink too. But one year, Cus was broke. He didn't have a dime. He said, "What am I gonna do? I have to give the kids a party." Just then, someone showed up and gave Cus $100. Cus had loaned the guy the money and he came by to pay it back! Since he was dead broke, Cus could've spent the money on other things for himself, but he spent it on a party for the kids.'

Graziano was one of those kids. For his moral health at least, maybe he should have stuck with D'Amato, rather than running with his new friends. He went to one of the same Catholic reform schools as Frankie Carbo and ended up in prison, where he met Frankie Peppo. This Frankie was a connected fight face who said to him the day he finished his stretch, 'Look me up on the outside, Rocky, you won't regret it.' So he went to see Peppo when he got out. Frankie didn't show – but the manager Irving Cohen and the trainer Whitey Bimstein did. They would be his handlers for the rest of his career.

And who managed and trained Billy Graham? Cohen and Bimstein. Billy didn't get to meet Peppo, either. He didn't have to. But years later, he would have cause to wonder about who was really in charge of his ring destiny.

Billy didn't resent the Rock's progress. Not at all. He wasn't that sort of a guy. He was straight up, a good friend and totally loyal, they said.

Billy screamed as loudly as anyone when Rocky knocked out Freddie 'Red' Cochrane that August night of '45. Cochrane had been around forever. He'd been good, very good. But he was old now. The seventy fights he'd put his body through had finally taken the stuffing out of him. Red couldn't punch a bus ticket. But he gave everything he had every time he fought, all the way back to the Depression years. He'd even got a decision over on old Jack 'Kid' Berg in 1938 when Britain's finest was a shadow of the champion he'd been. But to Red it was a big, big win, a name.

Mostly, though, the opponents that followed would not be that well known outside their neighbourhood. Many were rank novices. Against stiffs and downright no-hopers, Cochrane had worked his way up the rankings – and, against the odds, against all expectations, he'd been given a crack at Fritzie Zivic for the world welterweight title. In the unlikely setting of Ruppert Stadium, Newark, New Jersey, on 29 July 1941, he got an equally unlikely verdict from the one scoring judge, who also happened to be the referee, Joe Mangold.

Cochrane, technically, was the world welterweight champion and kept the title throughout the war – but, really, the belt was mothballed for the duration and Red kept his career going with non-title fights.

So, that's where Red was at when he fronted up in '45 to Graziano, who was good-looking, streetwise, a sassy fighter with a hell of a whack and moving up in the boxing world. And up a division. To the top of the middleweight pile. The Cochrane gate was good, $100,469, shelled out by 18,071 fans hungry for a thrill, and Rocky gave it to his roaring constituency in spades.

Graziano knocked Red down for an incredible seven nine-counts before putting him away in the tenth of ten. The place went crazy – and so did the papers. This was the bout that inspired the publicists to tag Graziano boxing's new 'Million Dollar Baby'. It would be a rough ride for him.

He had brought his previous outing with Cochrane in the Garden, just two months earlier, to another dramatic conclusion.

The *Ring* magazine voted it their fight of the year. It seemed every fight Rocky had, his legend grew.

Red? He just wanted one more big night. One night back in the Garden. It helped coming from Elizabeth, a town full of 'the connected' in boxing. Improbably, then, even though Graziano beat the living daylights out of him twice in as many months, Red got another farewell shot.

Marty Servo knocked Red out in the fourth round. It was in the Garden on 1 February 1946. Red reckoned he had no more to give. He went back to his dressing room, slung his gloves and kit in a bag and walked out, never to return. He'd just about beaten the system. And there weren't many fighters who could say that.

Rocky, meanwhile, was riding high. He had New York in the palm of his hand, with his big heart and big punch. The Rock, when it came down to it, could really hit. Billy could too – but he left his big punch in the gym, which was a mystery to his friends.

One of the gym rats and boxing faces, a year or two older than Billy and the Rock, who was hanging around the Garden at the time was Gil Clancy. He'd end up matchmaking for them (as well as managing a lot of valuable boxing beef, including Joe Frazier).

'My friend's mother was the top cleaning lady in the Garden,' he told me, 'so she used to get us free tickets for all the big matches. I've seen a lot.'

Clancy would see Graham in the gym, and one thing always puzzled him. 'Billy was a great fighter. The funny thing was very few people know that, in the gymnasium, he was a tremendous puncher. Once he got into the ring, though, he really couldn't punch. I have no idea why. I couldn't figure it out then, and I'm none the wiser now, fifty years later.'

At this stage of his career, Graham was marking time as Rocky was forging ahead. On 27 August, outdoors at the Queensboro Arena in Long Island City, he stopped no-name Donnie Maes in one – and then came defeat. Two weeks later in the same arena, on a night marred by storms, Graham dropped a split decision to Tony Pellone, a New Yorker who also boxed as Jimmy Pell. It was a sickening blow for Billy. He got his career back in the groove soon enough and he figured he had the beating of Pellone – yet a year later he would lose to him again, this time at the Garden amid howls of derision from his fans. One official saw it six rounds to Graham, three to Pellone, with one even. The other scores were 5–3–2 and 5–4–1 for Pellone/Pell.

'Don't worry, Billy,' Uncle Mike told him when he turned up next day to collect his cheque. 'They still love ya. You start winning, you'll get your shot. Trust me.'

Graziano, meanwhile, had sent out the message: he was an excitement machine. In 1946, he went on to figure in yet another Fight of the Year, KO'd by Chicago strongman Tony Zale.

That summer, Gavilan made it to New York, with his wet-behind-the-ears trainer Fernandez and enthusiastic manager Balido in tow. He'd earned nearly $7,000 in Cuba in the previous year and was ready to step into the American big time.

As luck would have it, when they docked, they met the renowned Cuban sports writer Jess Losada, who knew everyone not worth knowing in the boxing game. Losada took the wide-eyed trio down to Stillman's and introduced the Kid to Nick Florio, who had trained his esteemed compatriot Kid Chocolate.

Florio got Gavilan to go through his paces, gave him a few critical tips and said, yes, there was a chance he might make it. Florio was a boxing thoroughbred. He trained the heavyweight Roland LaStarza, Jersey Joe Walcott and many others; in years to come he would be in the corner of Floyd Patterson, Tony Canzoneri and, as a cuts man, Muhammad Ali. Gavilan, it seemed, had fallen on his feet.

However, on that day, in another corner of the gym, Losada was introducing Balido to one Angel Lopez. Señor Lopez was a big man in the Latino community in New York, the proprietor of the famed Havana-Madrid Cabaret and a sometime boxing manager, although he had no obvious credentials for the job – apart from his association with the man whose word in the business was usually final, Frankie Carbo.

This was to prove an association far more significant than that between Gavilan and Florio. The Cuban's life was to change dramatically in the years to come.

As was Graziano's, but far sooner. In 1947, the Rock's bad habits and connections caught up with him, like a killer left hook.

He recovered from the Zale beating and had a good Christmas at home. By now he'd moved back to Brooklyn – 1357 Ocean Parkway – and was living it up, just like Sugar Ray. Rocky was one hip man about town.

On a January Saturday afternoon, he was driving around Brooklyn in his shiny new Cadillac when he was stopped and taken in for questioning by the assistant district attorney, Alfred J. Scotti. The assistant DA wanted to know why Rocky had called off his projected comeback fight at the Garden against Reuben 'Cowboy' Shank. Why he was fighting him at all might have been a more interesting line of inquiry.

Shank, from Colorado, was managed by Chris Dundee, who knew his way around the underworld, and was the older brother of Angelo, who would go on to be famous in the considerable shadow of Muhammad Ali. Shank had shock wins over Fritzie Zivic and Henry Armstrong on his record, but he'd started to lose, to Sugar Ray Robinson among others. He would lose eleven and draw one of his last dozen fights in a career notable for his availability rather than his ability, and he had no right, really, to hope for a shot at Graziano.

But the word was out that Rocky was going to throw the fight against the Cowboy. So he pulled out, with a sore shoulder. Mr Scotti wasn't convinced. Was it, he asked, because he'd been offered $100,000 to lose to Shank and didn't want to go through with the fight? The Rock held firm under questioning. Sure he'd been offered a bribe, at Stillman's gym one day, by a guy he'd never met before, but he turned it down. He couldn't fight because he wasn't fit. No mystery.

After a month's deliberations, the DA and the boxing commission decided Graziano had broken rule 64 of the commission's laws, which said fighters were obliged to report bribes. They took his New York licence away – but he could still box in other states and, that July, he knocked Zale out in their rematch in Chicago.

Zale won the rubber match in '48 – but in the post-war buzz around New York, Rocky Graziano was still the man. His fans didn't care if he was mobbed up or not. He delivered. He filled the Garden and he thrilled the crowds. Billy Graham, meanwhile, was the always respected, often ignored consummate professional. When his chance came, he would learn how much connections count.

The guys who made the rules were watching Billy . . . and they kept a close eye on another fighter from Billy's part of town, Jake La Motta.

About this time, the district attorney's office in New York had just about had enough with the fight game. Even by the lax standards of the New York State Athletic Commission, professional boxing and the clowns who ran it had been taking liberties. The excesses of the thirties had not disappeared with the passage of the war; they had just been handed over to new people.

And La Motta was very much one of those people. He always protested he had little or nothing to do with the Mob. It was Jake's straight-faced opinion, whenever asked about them, that they existed in the imagination of newspaper guys, and even if they came calling, he would have nothing to do with them.

There is another myth at large that, at this stage of his career, La Motta was just a struggling middleweight waiting in the queue, albeit one with a great chin and big heart and an almost unquenchable thirst for fighting. But he'd already made a lot of money from boxing. He was a flat-nosed, ticket-selling guinea from the Bronx, who knew every face in the business. How was it he couldn't get a shot at the title?

At this point, La Motta had been in the game since March 1941, six and a half years. He'd had seventy-eight fights. And he'd made plenty of money. Enough to invest in two boxing clubs, the Jerome Stadium, across the street from Yankee Stadium, and the Park Arena, in the east Bronx.

Of course, Jake was not the owner of record of these establishments. It was thought a conflict of interest for a boxer to own the places in which he and his colleagues might belt each other senseless. The clubs, according to the taxman, were the property of a retired pug called . . . Joey La Motta, Jake's little brother.

On the night of Tuesday 11 November 1947, Armistice Day, Jake was at the Park Arena, pretending not to be the governor but shaking hands nonetheless as punters streamed in to watch what he could not declare was actually one of his shows. Top of the bill was a local lightweight called Frankie – sometimes Freddie – Palermo, who got a unanimous decision over Jimmy Pierce in an eight-rounder. Frankie was a crowd-pleaser without much of a punch, a guy who could take it and give it. This was his ninth appearance at the Park in twenty-seven outings going back to the end of the war. He'd made it to the Garden a few times, on the undercard.

But this night was significant not for the boxing but for the clientele. Jake's special guests, who arrived at the run-down ghetto in chauffeured comfort and safety, were Blinky Palermo (no relation – although who could be sure?) and Frankie Carbo. This was boxing royalty, two of the game's rarely seen elite. Their habitat was more often the dining room of the Forrest Hotel on 49th Street, where they would eat and talk with Mike Jacobs from across the road at the Garden.

About this time, though, Uncle Mike was not always on call; he'd had a cerebral haemorrhage in 1946, so taking care of business for him and the Twentieth Century Sporting Club was his lawyer and cousin, Sol Strauss. A couple of months before that Armistice Day of '47, Strauss had occasion to talk with Blinky about an upcoming fight at the Garden, between La Motta and Palermo's seemingly tough prospect, Billy 'Blackjack' Fox, a lean young hitter from Philadelphia who'd knocked out forty-two of forty-three mostly anonymous opponents, losing the one serious examination of his credentials, to the world light-heavyweight champion Gus Lesnevich by TKO that February.

The match was made for Friday 14 November. Fox would get 20 per cent of the gate, Jake 30 per cent. It seemed a straightforward piece of business, two top-class fighters headlining a big bill at the Garden.

Now, here they were, three days before the scheduled meeting of Fox and La Motta, Blinky and Frankie, in this dump watching Blinky's namesake, a second-rate pug, go against Jimmy 'Irish' Pierce, a Canadian of little ambition who'd been knocking around the rings of New York for seven years. This would be Jimmy's thirty-seventh loss from sixty-four starts, and he would be retired within another four. Or maybe they were there to see the rising heavyweight, Roland LaStarza? He'd been boxing just four months and this would be his ninth straight win, a perfunctory points verdict over an undistinguished New Yorker called Lorne McCarthy. LaStarza was twenty, good-looking and Italian. Within three years he would be giving Rocky Marciano a serious argument over ten rounds at the top of the bill in the Garden. For now, though, Roland and all the other guys were making mere background noise for Blinky and Frankie.

They ghosted into the arena, shook Jake's hand and, ignoring the undercard action, retreated to a quiet corner to talk. With them were his brother Joey, the brilliantly named promoter 'Honest' Bill

Daley and a PR flack called Billy Stevens, one of the fight writers happy to pick up the envelope left for him in his press seat.

As the gutsy New York boxing writer Barney Nagler observed, 'They were gaudily open-handed. The mobsters in the business usually were as cozy as dope pushers about their activities in boxing, but Carbo and Palermo had chosen to consort with Jake La Motta in a public fight club on a fight night, when officials of the New York State Athletic Commission were near enough to smell them.'

Three nights later, Friday 14 November, La Motta did what he always said he would never do: he threw a fight. He threw it so badly even he was embarrassed . . .

This is how the referee Frank Fullam, an apparently trusting man, described the fight in his testimony to the New York State Athletic Commission on Monday 17 November:

> In the fourth round when I stopped the fight, it was one-sided from the start. Fox rushed La Motta all over the ring, throwing both left and right hooks to the body and head without any return from La Motta. I only stopped the contest after it was evident to me that La Motta was unable to defend himself. The contest was stopped in mid-ring. I also felt that any further punishement might bring serious injury to La Motta.
>
> I want to state that during the course of the three rounds and two minutes and twenty-six seconds in the fourth round when I stopped the contest, during that time La Motta could have gone down from several hard blows he received from Fox, but refused to do so especially when he was in the corner and one of his knees nearly touched the canvas. As a referee of many contests, I thought that La Motta showed great durability to survive the punishment that he took. If I didn't stop the fight, it would have been a

> tragedy and perhaps the fellow might have been killed. As a representative of the commission, I am there to see that it is an honest, fair fight and to see that the boys are not hurt.

Fullam had been a good middleweight himself, in the thirties, and boxed four times in the Garden. It was not as if he didn't know his way around a boxing ring.

So, quite how La Motta thought a laboured performance against Fox would go unnoticed by everyone, including the expert Mr Fullam, is beyond comprehension. In fact, he didn't think that at all. As he wrote years later in his entertaining autobiography, 'If there was anybody in the Garden who didn't know what was happening, he must have been dead drunk.'

What the drunks might have missed was La Motta pretending to be backed up by Fox and faking it big time on the ropes.

The Bronx Bull did not so much as raise his horns once in twelve minutes of boxing against Fox. He had them booing from the stalls and the celebrity seats. In one stinking evening, he ruined what reputation he had. There wasn't a washed-up pug in the building who didn't know Jake had thrown the fight.

Not everyone saw it that way.

'There was no questioning Fox's victory and the dynamite which the Philadelphia negro packed in his fists,' reported the United Press, ringside at a hushed Garden.

But the word had been out for some time. The day before, bookmakers had them each at 6–5, a reflection, perhaps, of Jake's great chin and Billy's impressive punching power. By fight time, however, Fox was an 11–5 on favourite. Three hours before they got in the ring, the bookmakers had pushed him into 3–1 on, and stopped taking bets. The only money left in the house was that of dumb-ass fools still betting on Jake.

When Jake followed the referee into the witness box, he told the commissioners he'd boxed below par because he'd suffered a spleen injury in training. The following day it was Joey's turn to lie.

Yes, he'd heard rumours of a fix, 'all the time. All day people came over to me, and different people asked me, "There's a rumour going around your brother is going to throw the fight." I said, "My brother never threw a fight. He never intends to throw a fight."'

Then Palermo gave 'evidence'. No, he said, he had not seen La Motta at the Park Arena three nights before the fight. In fact, he couldn't remember if he'd been to the Park Arena or not. 'I have been in quite a few different places,' he said, without adding that a lot of those down the years had bars on the windows.

Other witnesses included Al Silvani, Jake's sometime trainer, an actor and Frank Sinatra's masseur. He hadn't heard anything either.

This investigation was going nowhere. So the DA got involved again. He called in Palermo and Carbo to his office in Leonard Street. They told him nothing. The assistant DA, Alfred Scotti, also questioned La Motta. Jake denied he'd ever borrowed money from Carbo. He did admit putting $10,000 of his own money into the Park Arena, but said he took it from his ring earnings, which he never liked to put in a bank, preferring to look after the cash himself in his apartment near Yankee Stadium.

The upshot was this: Jake was barred from boxing in New York for seven months and was fined $1,000 of his $20,000 purse. And that was that.

La Motta went in to the fight with a 64–11–3 record and, most importantly, a promise of a shot at a world title. The first one mentioned was that belonging to Lesnevich. Fox got the chance instead and was stopped in one. Jake, meanwhile, hung out for a go at the middleweight title held by the remarkable Frenchman Marcel Cerdan.

He got his fight with Cerdan, two years later in Detroit. Jake put the Frenchman over in round one, and the champ got up with a dead shoulder. He carried it for as long as he could, before retiring at the end of the ninth. And Jake, after all his sweat and conniving, was champion of the world. The Mob always delivered.

When La Motta finally confessed in 1960 that he had thrown his fight against Fox, he remarked with cynicism characteristic of those who live on the edge of morality: 'Dan Parker, the *Mirror* guy, said the next day that my performance was so bad he was surprised that Actors' Equity didn't picket the joint.'

This was it from the horse's mouth, in contravention of everything he, his associates and allegedly disinterested parties said at the time. Yet La Motta has been allowed to wallow in a peculiar brand of celebrity. Certainly, he was a great fighter but he was a hugely flawed individual when willingly swimming with those sharks. He had grown up with corruption, from the Lower East Side and reform school to the bright lights of the Garden, immortalised eventually by New York's most New York of New York actors, Robert De Niro. After Jake had quit, he could say what he wanted. 'You know, the way we were brought up and the place we were brought up, staying out of the Mob wasn't the easiest thing in the world.' La Motta, though, managed to skirt around the edge of gangsterism – after he'd paid his dues by losing to Fox.

Martin Scorsese, through the acting genius of De Niro, reinvented La Motta, like no other artist of illusion possibly could, in *Raging Bull*. This reinvention is telling. It represents boxing itself. It is the not-so-grand illusion. Fighting, where death is an option, could not be more real; the retailing of it, however, allows for all manner of invention. You see what you are encouraged to see.

One night in Manhattan, in an Irish bar near the Garden, I am talking to an old operator, an ever-laughing acquaintance called Bert Randolph Sugar, who has insinuated himself into every corner of the fight game since coming to New York from Philadelphia as an ad man in the fifties. He reinvented himself with fedora and cheroot as few would have the chutzpah to do, but he knows his stuff. Away from the many interviews he does, he lives the quietest of domestic existences, a Sunday-morning car-washer in less-than-loud clothes, a suburban husband and father with a good take on survival in a mad business. If anyone can spot bullshit in boxing, it is Bert Sugar.

I wondered what he made of La Motta and his contemporaries, the fighters and their handlers, the writers too, who had to get on with their work surrounded by unsmiling crooks with cold hearts and no real love of the sport.

He quotes an old fight writer, Marty Rivera (Mario Rivera Martino, who wrote for the *Ring* as long ago as 1945). He told Bert once, 'Yeah the fights were fixed, the fights were crooked in the old days, the Mob had them. But then you had better fights.' Sugar adds, gnomically, 'I mean it's not my observation, but I agree.'

And, as is commonly accepted, if you weren't in the mix, no matter how good you were you didn't get the big fights. Could this be why Billy Graham will be remembered as the best welterweight never to win a world title? 'Yes,' says Sugar. 'A lot of these guys, they weren't connected. But they could come in if they wanted and then they got connected.'

And, once you were in, the scam was the rematch, right?

'Yeah, but also it was a betting proposition. A fighter was making thirty thousand let's say, maybe twenty, [up front, according to the declared purse]. You'd give them fifteen or ten. You could lay off twice that. You pocket the difference. He's got more money and is told, "Let's go! You'll get another fight, Joe." I mean, when Jake did it, he didn't even tell his dad, who lost all his money on him.'

So, who were these men the betting naïf La Motta was mixing with? They were not choirboys.

At 10.45 p.m. on 20 June 1947, Eddie Cannizzaro, a low-life hoodlum with limited career prospects, put a single shot from his Carbine M1 through a window of a mansion at 810 North Linden

Drive, Beverly Hills, and drilled a bullet into the left eye of Benjamin 'Bugsy' Siegel.

Bugsy, whose final moments were spent reading the *Los Angeles Times* alongside an old friend, Allen Smiley, was the real godfather of modern-day Las Vegas. He was rendered unstable through drugs to the point where he thought he could double-cross the Mob, a fact latterly recognised by his girlfriend, Virginia Hill, who was holidaying in Europe at the time. It was her grand residence in which Bugsy died. He called Virginia Flamingo and, unreliable legend has it, named his Vegas casino after her. Virginia couldn't find time to go to Bugsy's funeral, though. She chose instead to call on her pal, the connected actor George Raft, whose asthma was playing up that day. You can have calendar congestion like that sometimes.

It wasn't until 1992, when Jimmy 'The Weasel' Frattiano sang for forgiveness from the FBI, that it was confirmed that the hit on Bugsy Siegel had been ordered by his old friend and killing partner from Murder Inc., Frankie Carbo.

Carbo – sophisticated, well dressed, charming and anonymous to all but his inner circle – was, they say, responsible for the deaths of five other men during his career as a professional life terminator for Murder Inc. That's what they say.

Those were the life-and-death crimes. Carbo's other sins were, according to people in the fight game, not so serious. Just business. Only fools believe otherwise. And there are enough fools in boxing to keep the business going for a very long time. Carbo's main killing field in boxing was Madison Square Garden. In 1947, he did not quite have a total grip on the place, as it was still in the possession of the Twentieth Century Sporting Club. He could wait.

Together, with the purblind acquiescence of other timid types, Carbo, Palermo and their friends would turn Madison Square Garden into a sordid hall of deceit. Or rather, they made sure it would remain that way. Because, in truth, it had rarely been anything else. For all the mythification and memorable nights, the Garden still stank with as much bad karma as it once did with liniment and cigar smoke. Frankie and Blinky, along with their sharp-suited cohort, Jim Norris, and the smooth-talking lawyer Truman Gibson, robbed boxers remorselessly within its walls and, from the East Coast to the West, they froze out good fighters who wouldn't do as they were told, arranged results and controlled titles – perhaps not so brazenly as some before them, maybe not so subtly as some that followed, but with no less enthusiasm.

And yet, remarkably, to this day people talk in reverential tones about the golden days, about how the mobsters might have been killers but they knew how to run boxing, what a sweet guy Carbo was, how Palermo never did them down, how Norris always kept his word, how fighters got good pay days.

As Al Certo remembers them, 'They were decent guys, they were men of the world. This is not your night, kid, next fight, don't worry about it, it's yours, shit like that.'

Al's from Jersey. He's been in boxing all his life, as a fighter and trainer. All his memories are good ones, because they are the only ones he wants.

'Like, for instance,' he says, 'a fight between Ernie Durando and Rocky Castellani [Madison Square Garden, January 1952]. He beat Ernie already, so this was a rematch. So on paper he would be the winner. It was a national television Friday-night fight and Ernie Durando was a tremendous fighter, he was my dear friend. He hit Castellani with an uppercut and he took the shot and both feet were off the ground. Down he went. Boom! It was a lucky shot, you know? It was in the back of the *Daily News* and you see Castellani, like . . . oh God. Anyhow, they stopped the fight and Castellani's manager [the gangster Tommy Ryan] jumps in the ring and starts throwing a million fucking punches. He hit everybody! He was the boss of the Mob, a big guy. The story is that he bet $80,000 on the fight. So it wasn't a fixed fight, right? It was a fight that shoulda gone all the way, but it didn't. Ernie got lucky. But they were the great years . . . and to fight in the old Garden, even a four-rounder, even a six, you fought a semi-final, hey, wow!'

Bill Cayton, who once owned a leg of Mike Tyson, said not long before he died that he'd heard of only one fixed fight in more than fifty years in boxing. Cayton, an apparently decent man who was properly stiffed by Tyson when Don King came along, must have been either deaf, blind or a fool – or all three. Gil Clancy shared Cayton's view.

Dan Parker, the most outspoken, incorruptible and determined of New York's boxing writers, pointed out the nods and winks many times in the fifties while more malleable colleagues, including the vaunted Nat Fleischer, long-time editor of the *Ring*, chose not to look too closely lest pesky questioning upset Carbo and his associates, among them the frightening Frank Costello.

What Parker knew, and where the cynics are partly right, was this: promoters and managers, in each other's pockets, didn't need to 'fix' that many fights; they only had to own most or, preferably, all the fighters, a lesson Don King learned well. A boxer was either in the loop, and got the fights, or wasn't and didn't.

'Evidence the New York State Crime Commission secured by wire tapping indicates that Carbo has a voice in the matchmaking decisions of the IBC,' Parker told his readers. 'He is supposed to have named the list to matchmakers at Madison Square Garden and has been known to give them orders. No one knows how many fighters the tight-lipped Carbo has a piece of or how often two Carbo warriors square off against each other in a "grudge" battle for the entertainment of televiewers – most of whom, incidentally, haven't the slightest inkling that such a person exists.'

Carbo's friends lurked to bad intent across a range of illegal activities. When they gathered at the Garden for the fights, they had plenty to talk about. There was a loose hierarchy and very much a common purpose.

Costello's patch was gambling and he had in tow on all the big nights Dandy Phil Kastel, who'd fly up from Miami, his appointed fiefdom, as well as the bookmaker Frankie Erikson. Also there much of the time, shaded and guarded, was Lucky Luciano, whose line was the numbers racket, heroin and prostitution. Nearby and unapproachable was Joey Adonis, who fixed elections and broke bones on the waterfront. Before they were offed or locked up were Siegel and Meyer Lansky, enforcers who freelanced in Philadelphia. Longy Zwillman ruled Jersey, the Mafia's American kingdom. They were just some of the faces who peered through the cigar fog at the moving meat in the ring, men most of them owned pieces of.

Idle suspicions? Well, Graziano and Sugar Ray Robinson both testified in the late forties that they'd been offered money to throw fights. So did La Motta – but only when the damage had been done and his legend had been cemented. Frank S. Hogan, a New York prosecutor not shy of a headline, had hounded down gamblers betting on rigged boxing and football matches in the forties. Isadore Dollinger, a Democrat from the Bronx, declared boxing fans had been 'duped, fleeced and otherwise damaged' by these shysters. Al 'Bummy' Davis, a tapped-up and very good welter-weight from Brownsville – later gunned down by four armed robbers while swinging bare fists in defence of his friends bar, Dudy's, in 1945 – was under constant surveillance. Saverio Freddie Fiducia, a Jersey heavyweight who also boxed as Freddie Martin, was suspended for 'laxity' in reporting a bribe to take a dive at the Garden against Freddie Schott, who stopped him in the ninth. And so it went.

These were not nods and winks. The sports writer Jimmy Cannon called boxing 'the swill barrel of sports'. And still most people looked the other way.

In 1948, the superb lightweight Ike Williams was told ten minutes before he was due to get in the ring with Freddie Dawson that he ought not to bother. The result had been pencilled in. Williams, devastated, went to the writers Red Smith, John Webster and Jack Sarnes and told them: 'After I fight tonight, come back, I'll have a story for you.' The writers shifted uneasily. They weren't sure about this guy and his story. Besides, who'd believe him? They collected their envelopes and looked the other way.

The officials got word of Williams's conversation with the reporters. The fix was off. The fight was allowed to run its natural course. Ike won a majority decision – but, later, he was fined $500 for implying the officials were corrupt. The chairman of the committee who handed down the fine was Leo Raines. Mr Raines was a friend of Carbo's right-hand man, Blinky Palermo. And all those writers Williams had tipped off knew that too. Did any of them write that? No.

As Gil Clancy told me, there was an unspoken bond between promoters and writers then. After big fights, they'd meet up at Toots Shor's. 'There were a lotta guys there most nights, guys like Al Buck, Lester Bromberg, all the top-notch boxing writers. I'd say so. I thought the best of 'em was Lester Bromberg. And you know, in those days, we used to pay the writers, the managers. After every fight, you'd have to have an envelope and give it to the writers. Yeah, we paid 'em. Under the table. And Lester Bromberg was the worst. Aw, he was the worst. You'd pay him after the fight, next day – you'd have a fighter by the name of Smith – and he'd write, "I thought Smith lost every round." I thought, what the hell am I givin' this guy money for? They all took the money, all of 'em. Regular envelopes. We'd have the envelopes all ready to give to 'em after the fight.'

Ike never stood a chance of getting his real story out there.

When John Garfield starred in *Body and Soul*, one of the more eloquent of many depictions of boxing's darker side, in 1949, there was righteous indignation reported in all quarters. One spluttering saint was Charles Johnston, the so-called president of the corrupt International Boxing Managers' Guild. He took exception to managers being painted as 'thieves, gangsters, fixers, connivers, double-crossers'.

Jeffrey Sammons, author of *Beyond the Ring: The Role of Boxing in American Society*, commented: 'If the film [*Body and Soul*] deserved criticism, it was for understating rather than exaggerating boxing's ills.'

This is how the triumvirate who controlled the Garden – Carbo, Palermo, Norris – preyed on the vulnerable, no matter what their standing. They did not have to beat fighters up or break their knuckles. They had merely to drop hints, sinister ones. They would let it be known a fighter risked more than merely losing a fight if he refused to follow instructions, or hand some of his contract over to them or their frontmen. And would you read about it in the papers of the time? Not likely. Lou Duva, who has been around boxing, and around New York and the Garden, for more than half a century, remembers the newspaper guys. 'I'll tell you one thing about the writers in those days. They were the best allies a promoter or fighter could have. They told in-depth stories. They had the reality, they built the fighter or they tore a fighter down, built him up or knocked him down but they were always out there doing their job.'

Boxers were the rest of the hired help in the fight game's ever-churning industry. A boxer who refused to cooperate would be left on the slagheap. Only the truly unreachable, fighters like Sugar Ray Robinson and Joe Louis, were considered immune. For Joe, however, it was an illusion too far.

8

ALONE ON SUGAR HILL

One of the saddest personal moments Joe Louis had to endure followed what is generally regarded as his poorest performance as champion.

Barney Nagler, a writer, fight fanatic and as New York as the smell of the summer subway, was there to chronicle the events before, during and after Joe's defence of the title against Jersey Joe Walcott at Madison Square Garden on 7 December 1947.

Five days later, Barney tells us, Joe was holed up in apartment 8E at 555 Edgecombe Avenue, on Sugar Hill in Harlem, spiritually adrift and physically wrecked. His face bore the marks of what ought to have been a defeat but, scandalously, was a result in his favour. Even Joe didn't think he'd won. After John Addie announced the verdict to a disbelieving Garden, winner Joe ambled over to loser Joe and whispered, 'I'm sorry.'

Later, the phone in the dark, noiseless apartment rang that eerie, noir New York telephone 'brring!'. Again and again. But Louis, essentially a solitary soul and as happy in his own company as when surrounded by admirers, wouldn't pick it up. Reporters were at the Garden, waiting to talk to the champ. Normally, he would be whisked into a room by Jacobs, the ubergrump of the Garden. Joe always liked Jacobs. In boxing, you take your friends where you can. Planted in front of the newspaper guys, Joe would go through his short-sentence routine, giving little away but the obvious.

Now, however, Uncle Mike was not so loud. He'd been struck down by a stroke the previous December, his right arm hanging limp, as it would for the rest of his life. Joe had split with Julian Black, because Julian refused to lend the fighter the money to settle his divorce, and the champ was easy prey. Except when watching the referee count to ten over a beaten foe, Joe was lousy with numbers. Now he employed a new adviser called Marshall Miles. He was unable to ward off a fresh set of wolves gathering around the champ, although he apparently meant well.

When Miles dropped by Edgecombe Avenue that December day after the fight to tell Joe he'd earned $75,968 for hanging on to his title, he saw a fighter in deep despair. Visitors came and went by the hour, unwelcome. Reporters kept ringing. All were rebuffed. Then, fatigued perhaps, Joe allowed just one up to see him, a local guy, from the *Bronx Home News*.

Nagler, naturally, wanted to talk about the fight. Joe wasn't keen. It was a sour memory. He went over to a dirty mirror and looked in. The bruises had not all gone down, his eyes were milky pink still, and not quite focused. But the most disturbing image coming back at him was of a champion who'd lost his way. Few knew it, but Joe was a broken man with money worries and an estranged wife who'd grown weary of his unreliable temperament. What sense of self old Joe ever had evaporated in that fight. He couldn't take in how he'd struggled against Jersey Joe. He wondered how he'd become so beatable.

Walcott wasn't a bad fighter. He was a smart mover, the inspiration, it is said, for Muhammad Ali and his shuffle. Jersey Joe was the first to blur his dancing feet like an illusionist, switching them mid-air as if they were blades on a grass-cutter, at the same time dipping his strong, brown shoulders, drawing out a limp and uncertain lead from his opponent before snaking through a counter and sliding out of range. On the night it mattered most, the Bomber's normally lethal right was stymied by age and Walcott's cleverly shifting torso.

Walcott reckoned that was his night, that cool Friday evening in the Garden. He'd waited so long, an unfashionable, smaller heavyweight with not much punch at the gate, just a sound, well-schooled challenger.

This was an age of style and swagger. Americans wanted invincible heroes they could gawp at, slaver over. They hailed Braddock because he was a battler too, but they treasured Louis because he was a god. The night before Louis–Walcott, Marlon Brando opened on Broadway in *A Streetcar Named Desire*. Jersey Joe? He could have been clipping tickets on a streetcar for all the fuss he caused.

They'd come to see Joe Louis. He was always the banker for the Garden. Amazing Joe v Ordinary Joe. No contest, reckoned the fancy, who, disregarding all evidence of decline in their champion, piled through the Garden doors to see him deal with this 20–1 underdog from nowhere. Walcott, whatever his pedigree, would surely go the way of the bums before him.

Louis hadn't fought in New York in more than a year, so appetites were whetted for a slaughter, one of those clinical, one-punch finishes only the champ could produce. It didn't matter to the crowds who the victim was, even if the promoters were concerned at the paucity of credible challengers, which allowed Louis to slumber on, fightless, playing a little golf, begging his wife to get back with him. Joe was getting dangerously complacent, they reckoned. He needed to earn his keep, to respect the title. Soon, they figured, they might run out of mugs to watch Joe beat up selected opponents.

So, after a wide trawl, they came up with one of Joe's former sparring partners, a guy from the past who'd helped him get ready for Schmeling in 1936. Louis, always tough on the hired help, had pounded remorselessly on Walcott in training eleven years earlier and had little regard for him now. It was his way. Not arrogant. Just getting the job done. If he was right, punters might be better advised to save their pennies for Tennessee Williams and the new kid on Broadway, snarlin' Marlon Brando.

Walcott's real name was Arnold Cream. He came from Merchantville, a small town in New Jersey, the son of a West Indian father who loved boxing and all its trappings. Joe's dad told his son about the wondrous welterweight from Barbados, the original Joe Walcott (who, even though hugely outweighed, once fought Jack Johnson), and young Arnold did his best to please his father.

As with Archie Moore, Sonny Liston and many other fighters born pre-war, Walcott's background was the subject of much speculation. His registered date of birth is 21 January 1914, but the *Ring Record Book* of 1960 says, 'Age has always been disputed, claims being made that he is several years older.' According to normally reliable sources, he started as a middleweight in 1930, taking fights anywhere they were offered, and soon he tired of being Arnold Cream. Who wouldn't? He hooked up briefly with the champion's people, 'Chappie' Blackburn and John Roxborough, and became Jersey Joe Walcott, but no closer to immortality. He went down with typhoid, got dumped and missed a year's money.

Thereafter, he stuttered through an undistinguished list of engagements. Smart but not big, dangerous but not particularly marketable, he received few favours from promoters or judges – at least in the early days. Walcott then was just another pug with slick moves and no connections. He fought part-time, going down to the docks, just like Braddock, to put a living wage together.

But Joe's life changed considerably for the better when a Jersey face from the gambling and liquor world called Felix Bocchicchio took him on after Joe Louis's connections had tired of him. At thirty, he got a second crack at making it as a fighter. Felix, luckily for Walcott, was connected. He was known around Camden, New Jersey, as Man o' War, and not solely after the famous racehorse of that name; Felix caused trouble wherever he went. He had an impressively long list of arrests on his manor in the thirties for the usual crimes and misdemeanours of his ilk: theft, jailbreaks, murder. Not many of the accusations stuck.

When Bocchicchio first saw Walcott he saw a meal ticket. He reckoned here was a heavyweight with a good jaw and heavy punch, a combination that would keep him 'live' in the business. At first Jersey Joe wasn't sure. He told his new admirer, 'Fighting never got me nothin' before, and all I want now is a steady job so my wife and kids can eat regular. I'm over thirty and just plain tired of it all.'

There were hundreds, probably thousands, of fighters like Walcott – good, experienced pros who knew their trade but who'd never had a break, never made the money they reckoned was their due. Years of slaving for small purses in front of indifferent audiences against other anonymous triers drained Walcott's spirit. He'd been Louis's sparring partner and, in the deeper recesses of his mind, that was his appointed place in the boxing hierarchy. Bocchicchio thought otherwise. He persuaded Walcott to give it one last try. He knew some people who might make it work.

He put his name around among the fight writers. He got him the fights against good opponents, and Joe kept winning. Soon, he had a rating and was within sight of a shot at the title.

A sports writer of the time observed that Felix 'knew as much about boxing as the Mona Lisa did about swatting flies, but he decided to learn. He was seldom seen without a fight promoter, trainer or prizefighter in his company. He learned from them the mannerisms of the fight game.'

A win over the rated Jimmy Bivins set a minor blaze under Walcott's career, and he made it to the Garden, finally, beating Lee Oma and Tommy Gomez, two not-bad operators. He also got the better of Joey Maxim, another name. All of a sudden, his trainer, Dan Florio, was encouraged to start dreaming dreams for Joe. He reckoned his guy might dazzle the old, fading champion, who was thirty-three and, as all the whispers said, looking lazy.

None of which convinced the bookmakers or the newspaper guys. Walcott, they said, didn't have a prayer.

'What went wrong, Joe?' the guy from the *Bronx Home News* wanted to know five days later. Everything, it seemed. Walcott beat Louis to the punch, slid in and out, clipped him at will around leaden jabs and unconvincing right hands – once his killer punch – puffed up his face and had the Garden booing their man at the end of an awful night for Louis. But, still, shamelessly, the announcer had to raise the old guy's hand. The referee, Ruby Goldstein, gave it to Walcott by a round; Frank Forbes had Louis two rounds ahead; and the second judge Marty Monroe saw Louis getting home by four rounds, a scandalous discrepancy. Walcott? He'd seen it all before. He came into the fight with a 44–12–2 log. He was there to make up the numbers. He smiled the grim, self-conscious smile of a man who knew he was always going to get dusted up.

'Man gets old,' Joe told Nagler in his unlit apartment. 'He don't take advantage of them things as fast as he used to.' He knew on the night (but didn't want to say now) that he shouldn't have got the verdict. Joe was back to deceiving himself. He was confused in many ways. His finances were in a diabolical mess. His credibility and reputation were in tatters. He was a paper king but he did not want to hear that. He wanted to drown out all the warnings Jack Johnson had given him before he died in '46. He wanted to listen only to the words of those loyal to his cause. And they were happy to dribble out the platitudes, like good court clowns do.

And then, as an act of relief, a salving of his conscience perhaps, he gave one young fight writer a scoop the big boys would have died for. He was going to fight Walcott one more time, set the record straight for all to see, and retire. He'd had enough. The *Bronx Home News* splashed the story across its front page and within hours the whole country, the whole world, was talking about the end of an era, the closing down of a boxing legend. But, if he thought his suffering was over, he was wrong. The fight game hadn't quite done with Joe yet.

In those years after the war, America was the 'can-do' capital of the world. 'In that era of general goodwill and expanding affluence,' says David Halberstam, 'few Americans doubted the essential goodness of their society.'

It was Joe's misfortune to be *in* that society but not *of* it – and to lose what judgement he had at the same time as his skills were dissipating beyond repair.

Now, in that quiet apartment on Sugar Hill, Joe gave the young boxing writer a scoop. 'I'm quitting, Barney. I've had enough. Print that if you want. I'm done with boxing.'

Among those reading the stunning news of Joe Louis's retirement plans was a handsome, dark-haired businessman in Chicago, a gambler and bon viveur who liked a drink, sharp suits and sharper friends. Jim Norris had star quality, an electric presence even in the rambunctious, playboy circles of sport and gangsterism in which he glided so effortlessly.

He liked to bet on anything that moved – horses, hockey players, fighters – and, thanks to his rich father, had a piece of several venues, including three of the six available ice-hockey franchises, that allowed him to so indulge himself. Maybe he got the bug from Norris Sr, because he'd been by his side from the time he was a small boy and he'd seen one of the great moments of fight-game history.

In 1919, on the fourth of July – when else? – at the Bay View Park Arena in Toledo, Ohio, James Norris Sr held the hand of James Norris Jr, aged twelve, and they sweated in the 120-degree heat alongside a heaving pit of pimps, gunslingers and gamblers to witness Jack Dempsey, straight from five KO wins in five months, take on Jess Willard, the ageing heavyweight champion of the world. As they entered the arena, Norris and his boy passed Bat Masterson and Wyatt Earp, who were collecting guns and knives from the congregation. The Norrises passed muster.

Young Jim's eyes remained wide open for the three rounds it lasted. The fight and the shenanigans that surrounded it in Toledo that day were to shape Norris's life.

Dempsey was twenty-four, lean, hungry and well connected. Willard was thirty-seven, and an odds-on favourite. Interestingly, his price drifted from a red-hot 5–17 to 4–6 to a lukewarm 4–5 as fight time approached. Somebody knew something. Ring Lardner, waxing more lyrically than usual, wrote beforehand, 'I do not care whether Dempsey win or lose, 'cause I got them there Toledo blues.' He did care, though. He had $500 on Willard and thought he was a certainty, but he also thought something was awry about the whole do.

Whatever the proper price, whatever had gone down, the champ seemed to lack fire leading up to the event. But he still had a farm to run. And he had a past. They couldn't rob him of his past. Willard, lumbered with the pantomime *nom de guerre* of the Pottawatomie Giant, once drove an opponent's jawbone into his brain with an uppercut. The boxer, John William 'Bull' Young, died the next day and Willard, a sensitive man, took a long time to recover from the incident, once he'd beaten a manslaughter rap.

Given the disparity in size and the slimness of the various odds, Jess went to the ring as yesterday's man but, according to some educated guessing, he was still dangerous.

Tex Rickard was the promoter – and one of two judges. He employed as his timekeeper an old 'associate', the former American and Canadian amateur heavyweight champion, Warren Barbour. Mr Barbour, who wouldn't turn pro, was considered above any suspicion. Twelve years later he would become a senator on behalf of the Republican Party in New Jersey. So he must have been a trustworthy citizen.

Rickard thought Dempsey might be killed, which would not look good on his CV. Young Jim Norris would be told later by his father that Doc Kearns, Dempsey's inventive manager, saw it differently. According to one story, the Doc had bet Jack's entire purse, $27,500, at 10–1 that the challenger would win in a round. Other versions – including that of Kearns himself in an interview many years later – had the bet as $10,000 to $100,000. Either might have been true. Like much that attended the fight, it has been coloured beyond recognition.

On the face of it, which is not always a reliable place to look for guidance in the fight game, Rickard had fair reason to believe it was a mismatch. Willard entered the ring at just over 6ft 6in, with a reach of 83in, and towered over Dempsey. Only Primo Carnera, Vitaly Klitschko, Lennox Lewis and Nikolas Valuev in the history of heavyweight title holders were bigger.

What happened on 4 July was Dempsey, from the first bell to the last, gave Willard four stone and the most sustained and brutal beating anyone had seen in a prizefight since the bare-knuckle pomp of John L. Sullivan. Willard went down seven times in the first round, his jawbone shattered in several places. The referee appointed by Rickard was Ollie Pepcord, a nervous man who was officiating in his first title fight. Ollie chewed a lot of gum. Near the end of the first round, after what Jim Norris and his boy thought was the final knock-down, he raised Dempsey's hand. Dempsey and his entourage stepped through the ropes to wallow for a moment in the reflected glow of the new champion, then headed towards the shade. Kearns immediately began celebrating his sound judgement. What a wise owl he was. He'd done it again, clipped the vigorish and outsmarted the smart guys – at which point Rickard's timekeeper, the upright Mr Barbour, cut everyone short. 'The count only reached seven!' he shouted at the ref.

Jack and the Doc, somewhat stunned, were recalled to the ring to finish the formalities of a near execution. Dempsey did what he had to do and, after two more rounds of unnecessary carnage, they carried Pottawatomie's finest from the ring like a sweat-sodden mattress. The crowd, intoxicated on violence and beer, screamed 'Quitter!' at the poor man. Jess couldn't hear them. As well as a broken jaw, he took away as souvenirs two broken ribs, six cracked teeth and a ruptured eardrum. The big, brave, delirious loser mumbled through ripped lips, 'I have $100,000 and a farm in Kansas . . . I have $100,000 and a farm in Kansas . . .'

They say Jack's gloves were loaded. Kearns even admitted as much years later, although such was his reputation for general tomfoolery, not many believed him, least of all Nat Fleischer, imperious editor of the *Ring* and the supposed gatekeeper of the truth in boxing. Young Jim Norris read that article and wondered if the truth was ever allowed to impinge on the running of prize-fighting.

What he came to realise was that it was a performance, captured on film, that established Dempsey's legend and did more than any other fight of that time to make professional boxing the number-one attraction in American sport. This was a business worth engaging.

The undeniable sporting fact is Jess Willard was the bravest losing champion in all of heavyweight boxing. His courage that day was perhaps matched in the heavyweight ranks only by that of the Frenchman Georges Carpentier in the tenth round of his fight with the champion Gene Tunney, five years later at the Polo Grounds in New York.

Even if the Dempsey–Willard fight read like a bad dime novel, it would be nice to think chunks of the story were true. Anyway, Dempsey was raised instantly to a standing somewhere above the president. And, after all, who'd want to disillusion a twelve-year-old boy at his first prizefight?

Young Jim couldn't sleep that night, before they headed home to Chicago on the train. He wanted more days like that. An ordinary life was not for him. The excitement of mixing with Kearns, Rickard, Dempsey and Willard, not to mention Wyatt Earp and Bat Masterson, stayed with him all his days. From that moment on, Jim wasn't fussed about the company he had to keep in pursuit of a thrill. When he grew up and inherited the family business, Norris did not baulk at choosing dubious company as he looked for glamour and kicks. It was not that he saw good in every man, just that he did not care if he saw bad. He could handle it. He'd seen it all in Toledo. He rushed towards bad. He embraced bad and, before long, bad became his drug of choice. Much later, when he was overseeing what Barney Nagler characterised as 'the decline of boxing', Jim Norris would do serious business with some of these men, notably the seemingly indestructible Doc Kearns.

'Jim Norris wasn't a bad man,' Truman Gibson told me a few years ago. 'He was just a gambler. He got a kick out of hanging around those guys.'

Thirty years after watching the great Dempsey, Norris, now a moneyed entrepreneur in Detroit, Chicago and New York, indulged his passion for thrill-seeking without a care about right or wrong. It was 1949 and New York was alive. Jim's decisions and actions would have a major impact in the city for a decade to come, as well as on the direction of professional boxing. He would be drawn into a web from which it was impossible to claw clear, but he didn't care. He didn't care much, either, what the papers said about him. Generally, the writers loved Norris. Given the company he began to keep, dandy Jim was not naive enough to think his press would be favourable forever, though.

His friends were mobsters and gamblers, liquor sellers and casino owners, fighters, managers, promoters. Just the sort of glamorous villains he'd first seen back in Toledo in 1919. You can imagine some of the less discerning among his new chums were delighted to have such a rich and amiable associate to play with. You can only wonder how Frankie Carbo must have laughed when he told his pal Blinky Palermo that tying up the world heavyweight championship and everything else at Madison Square Garden was going to be a neat wheeze indeed.

So, Jim Norris – and more pertinently to our story, the International Boxing Club that became his tool – got into the big-time heavyweight scene in a major way. As the boys on Jacobs Beach knew, Uncle Mike was a busted flush. His hard-of-hearing sidekick Sol Strauss had none of his clout or smarts and moving in on the Garden would be an easy gig for the IBC. Total control of the fight game was not far away for Jim Norris and his friends.

In his hotel spa in Hot Springs, Arkansas, Owney Madden, worth a modest $3 million and untouchable, put his feet up and tuned into the news from New York. He'd got out of the fight game when the Feds kept at him for giving house space to the likes of Lucky Luciano. He'd had to surrender his British passport in 1942, when they threatened to deport him. Finally, he was a paid-up Yank, with a TV set and a Frigidaire. Now he could knock the top off a beer, flick on his television and watch how his gangster associates were carrying on his work. He took a sip, balanced the beer on his belly and let the ambiguous grin of a contented villain spread lightly across his smug chops.

Joe's nightmare performance against Jersey Joe Walcott was the beginning of the slide. Truman Gibson was with him now and in a position to bring his legal and business brain to bear on what was becoming the dismantling of a proud champion.

But their next venture together illustrates how Gibson had the priceless gift of seeing only what he wanted to see. Louis Greenberg and Johnny Roberts owned a Chicago brewery and approached Joe to join them, along with Sugar Ray Robinson and a friend of Joe's called Freddy Guinyard. At first it went well – until Greenberg, who had worked as a 'treasurer' for a bootlegging operation run by Al Capone, persuaded them to open up in New York. It didn't do well, and Joe's reputation took a big knock when Greenberg's picture was published in *Life* magazine alongside Capone's. These were elemental errors of judgement, by Louis and by his attorney.

Joe, meanwhile, decided to fight on. He had the taxman on his back and, more to the point, nothing else to do. The self-deception was now deeply entrenched. Gibson did not feel disposed to argue with the champ and continued to do what he considered his best for him, setting up a company to look after his money, called Joe Louis Enterprises.

How Gibson characterises the next phase of their relationship is an interesting exercise in glancing back fondly on dirty waters. It is true you could search forever in the fight game for a virgin and find only whores and pimps, but Gibson really did look the other way at the most unfortunate times, perhaps assuming the best in people, when an intelligent man, such as he undoubtedly was, must have suspected these comedians were at the least not to be trusted with the chump change.

Gibson, wary of the influence of the ailing Jacobs and Strauss, had drawn his troops back to Chicago, where he felt more comfortable, and thought the future was opening up nicely when, one day, he took a call from the publisher of Hearst's *Herald-American*, George DeWitt.

DeWitt reminded Gibson of Hearst's Milk Fund, supposedly a charity to give milk to underprivileged children, but actually a front for other funds in the empire, most notably money generated in boxing. It gave new meaning to the expression 'milking it'.

DeWitt introduced Gibson to Harry Voiler, a member of Al Capone's gang, a Hollywood character who had been Mae West's road manager and was later jailed for stealing her diamonds. He'd also strong-armed for the Hearst organisation in a newspaper strike. According to DeWitt, Voiler was a citizen worth saving. To Hearst, that might well have been the case; quite why Gibson agreed with him is a mystery – unless you accept that Truman was a towering naïf.

Voiler offered Louis $250,000 for his contract, and would finance the deal by remortgaging the Mary Elizabeth Hotel, which he owned with his wife. Gibson went for the arrangement. However, when he, Louis, Marshall Miles and Joe's publicity man, Harry Mandel, went down to Miami to explore the options, Mrs Voiler nixed the deal.

So, rather than regroup, they went with plan B, inspired by Mandel. This was to prove a momentous twist of fate. The publicity flack suggested ringing Jim Norris, who was prominent in that part of the world as a lively social companion and toper, at his plush Coral Gables residence. So off they went, setting in train negotiations that would touch all their lives.

Joe was not the total fool some people around him thought. He was tired and he knew he'd been lucky to get past Walcott. Joe beat Walcott decisively in their rematch, knocking him out in the eleventh round in June of '48, this time at Yankee Stadium. He looked better. But not so much better the fans still believed in him.

He really wanted to quit now, while his reputation was vaguely intact, but he needed money. Norris, a keen student of the game since his trip to Toledo, was aware of Joe's degradation as a boxer. Not that you needed to be a genius to see it.

Gibson's plan was similar to the one Gould devised for Braddock when he won the title from Baer: he was going to rest the champ and let the challengers queue up in a series of fights that would create maximum interest in a bout some way down the road for Joe's title.

Gibson had come to the negotiating table well armed with

strategies. He had mapped out fights for the leading contenders, and had already signed up Walcott and Ezzard Charles. Norris was hooked. So was his partner, Arthur Wirtz, a Chicago real estate millionaire. 'After I got to know him,' Gibson writes, 'I found Wirtz to be the shrewdest and wisest businessman I had ever met.'

Wirtz promoted ice dancing, among other wheezes, and Joe, ever alert to the prospects of romance, wanted the Chicago man to introduce him to one of his skating stars, the Olympian Sonja Hennie. He and Norris agreed – and promised Joe $20,000 a year for his title. It was a grubby, unwholesome deal, horse-trading in flesh and cash of the lowest order. And no more than anyone would expect in boxing.

'Thus,' writes Gibson, 'was born the International Boxing Club, in March 1949.' There were a few further manoeuvrings to complete first, though.

Joe announced his retirement for a second time, on 1 March 1949. The IBC bought out Mike Jacobs, who quit the Garden to cope with his stroke and humiliation. Ned Irish, the managing director of the property, contacted Gibson and proposed a new operation – and that was, more accurately, when the International Boxing Club was born, with Norris and Gibson in charge.

It soon became evident that the tentacles of the deal Louis struck with these quasi-hoods would reach much further than the offices in Chicago and, later, New York. They would spread into the underworld of Carbo and Costello, a world that had no boundaries.

In June, under the auspices of the National Boxing Association, rivals to Mike Jacobs, Ezzard Charles beat Joe Walcott for the NBA version of the title. Louis defended his more widely recognised version of the title one more time, in Yankee Stadium that September, as he was always going to do, and lost it to the smooth-boxing Charles. It was dull and one-sided, Charles outpunching Louis two to one over fifteen rounds.

This was the drawn-out passing of an era. The Brown Bomber

could bomb no more. Uncle Mike was mute and poorly. The only constant was the silent menace of the men in the shadows. They knew, too, that to strengthen their grip on boxing they needed the Garden to remain the focal point of the industry. It was their power base. But they knew too that, with Louis a spent force, they could not count on fans flocking to the fights in the numbers they had done for two decades and more. They had to cover their investment.

Gibson told me years later the deal they struck with Louis 'saved the Garden'. It did – on the back of the new phenomenon in sports entertainment: television.

The key shift in the popular imagination of post-war Americans occurred sensually. Cheap entertainment until then had been at the movies and on the radio. Now it would switch from largely aural to visual, and it was delivered to the home. There was no need to go looking for a transitory fix in a cinema, although those pictures lingered with some vivacity.

Television was to revolutionise not only the way Americans consumed their entertainment but how they thought about life. It was to become the drug of the masses – which was not exactly how Herbert Hoover, then the Secretary of Commerce, envisaged its development back on 7 April 1927, at the AT&T Bell Telephone Laboratories auditorium in New York, when he told a distinguished audience, 'Today, gentlemen, we have, in a sense, the transmission of sight for the first time in the world's history. Human genius has now destroyed the impediment of distance in a new respect, and in a manner hitherto unknown.'

The occasion was America's first demonstration of television. In the flick of a switch, embracing the invention of a Scotsman far away, the New World got newer. No gadget, until the mobile phone and all the time-draining accessories attached to personal, portable computers, would so change the lives of ordinary people.

Hoover, though, would not be remembered for his brief and self-serving association with the dawn of television. Within two years he would be the 31st President of the United States. He swept to office on the back of just the sort of optimism he'd tapped into that day in New York. But his legacy would not be shiny new machines and unstoppable prosperity; it would be the tin and cardboard shanty towns across America that housed the human detritus of the Wall Street Crash. Then came the Great Depression – as unfortunate a rendering of the word 'great' as was the description of the 1914–18 World War. It was economic misjudgement on a grand scale and its consequences engulfed most of the privileged world all the way up to the beginnings of man's next 'great' cock-up: World War II.

In the thirties, then, the carefree and reckless optimism of the twenties gave way to the dispiriting hammer of realism. There were no televisions in Hoover's hovels.

Axiomatically, Hoover did not long hold the spirit of internationalism that television might have offered in different circumstances. The century's most dynamic economy turned back from technology, to the farms and plains. The president, a decent but intellectually feeble man (who years later, when Germany was dropping bombs on London, steadfastly opposed entering the war), acquiesced to Congress's demand for protectionism. So America retired behind tariffs that turned fields and dreams to dust. Exports were cut in half. One in four workers could not find a job. American insularity was cemented.

Twenty years on, the family focus was shifting swiftly from the radio to television. With the rise of TV, the hunger for instant satisfaction grew enormously. The year the war ended, there were only 7,000 TV sets in America, serviced by nine stations. Three of those stations were in New York City, two in Chicago and two in Los Angeles, as well as stations in Philadelphia and Schenectady, New York. The vast majority of the population still got their broadcast entertainment over the radio – but there was fascination, nonetheless, with the new gadget.

When Gimbel's Department Store in Philadelphia had a demonstration of a TV set, it attracted more than 25,000 people over three weeks. They were fascinated by programmes sent from NBC in New York, as well as by the local station. It was curiosity, but the word was spreading – and boxing, the sport that might have been invented for the screen, played a major part. Inevitably, Joe Louis was the attraction.

In June of 1946, NBC and Gillette billed their screening of the title fight between Louis and Billy Conn at Yankee Stadium as the first 'television sports extravaganza'. Ratings of 150,000 on 5,000 sets around New York might sound insignificant stacked up against the many millions across the world who watch fights today, but a year after the war it was a signpost moment in the sports entertainment industry. And it did not go unnoticed in Carbo's suite at the Forrest Hotel on 49th Street, just across from the Garden. It was there that Gibson, Norris and Palermo convened with Carbo to discuss how best they could exploit TV as the IBC settled down to business. Gibson and Norris were especially persuasive about the attractions and possibilities of the fledgling gadget. No longer did they have to rely solely on ticket sales to make a profit; they could negotiate contracts up front with the networks, who were desperate to provide quick, reliable entertainment for their advertisers. Sponsorship was paramount in the early days of American television as the makers of everything from razor blades to beer rushed to associate their names with a sport that still had a hold on the public imagination.

TV could not live on the sale of sets alone. Advertising, which New Yorkers had been addicted to since Madison Avenue was a dirt track, was to revolutionise the selling of the medium and the spread of televised boxing. In 1948, there were 993 registered sponsors of TV programmes, a staggering rise of 515 per cent on the previous year. This inspired a rush of applications for TV franchises across the country, 108 of them. The momentum was in place and would not die.

Television was still about entertainment, however, and would not have flourished were it not for the personalities who lit up American lives: Ed Sullivan (who would carry on broadcasting for an impressive twenty-three years, straddling several fads and styles), Milton Berle – who, at one point, had an unprecedented and never-to-be-matched 86.7 per cent of the available audience – George Burns and Gracie Allen, all of them old-style showbiz types. Youth had yet to have its day.

Much of TV was dull and parochial. The high spots were rare – which changed somewhat in 1949 with the first cross-country link, bringing together the best either side of the nation had to offer. That year, TV finally overtook the older medium and life would never be the same for Americans, whose love affair with television was now just about complete. A key moment of the transition from audio to visual, according to David Halberstam, was the axing of *The Fred Allen Show*, which had been an immovable radio staple in nearly every American home since 1931. Allen reacted bitterly to being dumped. TV, he scoffed, was 'a device that permits people who haven't anything to do to watch people who can't do anything'. He once observed acidly of Ed Sullivan, American TV's first talking head, 'Sullivan will stay on television as long as other people have talent.'

(In 1956, after it had time to bed down, A. J. Liebling, boxing's best literary friend, was still calling television 'a ridiculous gadget'. Free-to-air televising of his favourite sport, he told his army of readers, 'knocked out of business the hundreds of small-city and neighborhood boxing clubs where youngsters had a chance to learn their trade and journeymen to mature their skills'. It was a widely held view, especially among traditionalists.)

Allen, one of the great wits of the radio age, commented that 'young men with crewcuts were dragging TV cameras into the studios and crowding old radio actors out into the halls'. More pointedly, they also took with them the advertisers. *Variety* called it 'the greatest exhibition of mass hysteria in biz annals'.

In one sketch on his show, Allen co-opted Joe Louis to come on air and train him for a 'fight' with Jack Benny. They had conducted a bogus feud for years, Fred and Jack, but so seriously did listeners take his show, and Joe's presence on it, that schoolchildren gathered outside the broadcaster's house yelling at him to 'beat Benny for the good of Dorchester, Massachusetts'.

People believed in radio. It relied on their imagination. And there was what had always seemed an unbreakable bond between performer and listener; in 1950, there were 108 series that had been on air for more than a decade. TV required a different, lazier sort of faith. The images were indisputably there, on the screen. Listeners did not have to imagine Joe showing Fred how to throw a right cross. Nor did the words that so illuminated many brilliantly atmospheric radio broadcasts of big fights have to be so all-encompassing now. The pictures told the story, and what you saw just had to be true . . . didn't it?

'They were still innocent times,' Bert Sugar says. 'It was way before Watergate, way before Nixon, way before they shot JFK. America still believed in stuff. After that . . . we never really believed in anything ever again.'

There was no stopping TV in the fifties. Cinema viewing dropped by up to 40 per cent in cities that had television stations; fifty-five cinemas closed in New York in 1951. It probably started in 1947, when RCA made 170,000 seven-inch sets, and sold them all before Christmas. Two years later there were a million sets across America, then, in just twelve months, that number had grown to 10 million. One study suggests that in the early days American television showed as many as 120 Westerns. Americans tuned into *Gunsmoke* and *Wagon Train*, memorising all the catch-phrases. There were also Lucille Ball on *I Love Lucy*, the Nelsons on *Ozzie and Harriet*, wholesome entertainment of the old kind . . . and a whole range of fighters at Madison Square Garden.

NBC were already screening the *Gillette Friday Night Fights* from the Garden; Gibson now expanded the picture considerably and added shows, sponsored by Pabst beer, from Chicago Stadium, which he owned, on Wednesday nights. There were shows too from Miami, Los Angeles and San Francisco. The lawyer who had spent the first half of his career fighting for justice for his people was inextricably involved in the boxing business now. It was some career change.

Gibson used his contacts adroitly. He called up Bob Kintner, with whom he'd worked in the public relations bureau of the War Department and who now, fortuitously, was president of the American Broadcasting Company. Soon, ABC got in on the boxing scene, broadcasting fights on Saturday and Monday nights.

Matchmaking sounds such an innocent term, as if it were a mere bringing together of athletically compatible parties for their mutual benefit. In boxing, it hides more sinister truths. To be matched at the Garden, your manager had to be a fully paid-up signatory to an organisation misleadingly called the International Boxing Managers' Guild. It bore no relation whatsoever to internationalism, craft or a history of shared skills and everything to the strong-arm tactics of the Mob, most specifically Carbo. If you did not agree to the terms of the guild, if you did not sign with a manager from the guild, if your manager did not agree to parcel off bits of you to unseen, unlisted shadow investors, if you did not agree to fight when, where and how was convenient for others, you did not appear on television. If you did not make it on to TV, you were consigned to the outer limits, boxing in front of dwindling audiences in small venues.

Boxing was growing into an all-embracing octopus through the young and robust medium of television. Given the size of the receivers, no more than seven inches across the diagonal, boxing, conducted in a static ring, was far more amenable as a product than, say, baseball or football whose more extravagant movements across a field were harder to capture.

The development of the partnership between TV and boxing would have profound effects on the sport, from its roots to its apex. There was much money to be made by the few, and hard times ahead for those not drawn into the inner sanctum. That close-knit high command of boxing would tighten their grip on their sport to the point of near strangulation.

The IBC were paid $40,000 a week for their fights. Gibson says 60 per cent of that money went to the fighters and there was, indeed, a set scale. Main eventers at the Garden and in Chicago earned $7,000 each, fantastic money for the time outside world championship contests.

To feed the monster, the IBC still needed warm bodies. Before the war there had been many thousands of willing desperates, men like Braddock, a middling boxer happy to fight through all adversities to escape hunger and the dole queue. Once champion, he inspired thousands of kindred spirits among America's 15 million unemployed at the height of the Great Depression to keep going, against the odds. Later, although the war had robbed America of fewer young men it had in European countries, there was still a sense of emasculation after the conflict. Fighting men came home and began looking for something quieter than violence, sanctioned or otherwise. They now wanted to get their thrills outside the ring looking in.

Late-forties America was in a huge state of flux, confusion almost, as it searched for a new way to follow the turmoil of war. At the start of that decade, 8 million Americans (almost 10 per cent of the working population) could not find a job; by the end, unemployment had fallen by nearly 5 million. Life was considerably more pleasant as the fifties unfolded. More than 6 million Americans owned a car; the average salary was nearly $3,000 a year.

Not everyone thought television was a wonderful breakthrough. Gil Clancy, an old New Yorker and ring ancient whose voice still occasionally lights up a broadcast, is one who fell in line with Liebling and Fred Allen.

'Pro boxing, even when the Garden was the Garden, when it was on TV, Friday-night fights, they never got big crowds,' Clancy told me. 'Not at all. It was all because of television. You'd have a guy fightin' the main bout and it was a standard fee, he'd get $4,000 from television and a percentage of the main gate. Sometimes a percentage of the main gate didn't even come to a thousand dollars. You might get 3,000 people there because you could see it free on television. I'd say that was the case maybe around '52.'

Lou Duva, who was born in 1922 and has fought, trained, managed and promoted for nearly seventy years, in his native New Jersey and all parts of the world, knows that TV is the lifeblood of boxing. Yet he blames it for the sport's gradual decline in the fifties.

'I think it all changed right around television time,' he told me. 'They ruined the business.'

Perhaps 'they' did. But maybe people just tired of the fight game, the fixes, the lousy matches, old boxers hanging on too long, just because they were connected, young fighters never getting the breaks, just because they weren't.

Duva, one of boxing's most combative and straightforward characters, even as he gets towards the end of a fascinating career, doesn't put as much blame on the Mob as many do.

'You had to have connection,' he said. 'They'd come over and say I've got a part for you. There were good fighters around and if you wanted to go to the Garden they had the connection.If you wanted to go anywhere at all you had to have the connection. A few good fighters might have missed out because they wouldn't come to the party. But not many. They had a Managers' Guild at that time. It was a case of control and the guys that worked behind the guild were the guys that controlled it. If I was a manager and I had a fighter and I wanted to fight in the Garden, no matter how good the kid was, I had to join that guild. Mob guys controlled the guild, sure, and they usually were strong enough to make their guy the matchmaker in the Garden, stuff like that, or in any of their clubs. And I'll tell you something. As much as you say they were Mob guys or they were guys who had connections, in a way it was even better at that time. And I'll tell you why it was better. If you had a fighter and a guy came along and he started romancing the fighter and he's trying to steal the fighter away from you – you understand? – then, all of a sudden, the guy says I'm leaving you, I'm going with him. They had some recourse. They would go to that Managers' Guild and they would present their case and they could be called in by the right people and sit down and they'd say what's the problem over here, you know? Well, he wasn't getting me enough fights or enough money and this guy has promised me he's going to do better for me. Well, these people here would sit down and make sure the deal was done right. You want the guy? Either you buy him, or you work out some arrangement. And they always done it that way. Today you haven't got that. Like, if a fighter is fighting for a certain television station, he can't fight for another station.'

The truth, as ever, encompasses some of all those recollections. There were great nights, and there were ordinary ones too. What Duva describes as an amicable chat between fractious fighters and managers, settled over a gentlemanly meeting with the guild, was in many cases out-and-out intimidation. Fighters had their contracts sold on to connected faces, frontmen, knowing they had no alternative. They had no bargaining power but that which they held in their fists. And, if they didn't get promoted, they slid from view, down the bill and out on to the street, where only memories sustained them.

Eventually, when Walcott was used up and old, he turned into part of the nostalgia industry in boxing. He had wanted to get out, to get a proper job, but was talked out of it by the connected Felix Bocchicchio, and ended up briefly a world champion. But he did not carry much charisma with him. His time, glory and money passed. When he quit boxing he had little to trade on but his name. And, reveals Duva, there was one good man, often derided for his tightness – not to mention his alleged Mob connections – who helped him out.

'Rocky Marciano was my main guy,' he told me. 'He was tight with money, sure. He hired Jimmy Breslin, $50 a week, they did a TV show, *The Main Event* [the name of Duva's promotional company]. Rocky would turn up to the studio in old clothes so they'd give him a new suit every time. Ha! He was my personal friend but he was one of the cheapest guys. And let me tell you what he did for Jersey Joe Walcott. I'd seen him take care of guys like Joe. If you were running a dinner and you wanted to hire Rocky you'd pay him say 25 hundred dollars to make an appearance. But if he knew that one of the fighters, Joe for instance, was destitute he'd call back and say, hey, listen, you're paying me 35 hundred dollars now, something come up. And he'd give the money to Joe. That was Rocky. A great man.'

Rocky's time at the start of the fifties was about to arrive. He would become a phenomenon of his decade, a TV boxing star, one of the first.

It took a genuine star – La Motta, Robinson, Louis, Marciano – to fill a place. Anything but the best, any fight you didn't have to be at, you'd watch in the bar. Some fights, though, you had to be there, you had to be able to tell your kids you were there. They'd want to know why, otherwise.

Joe Louis still had more pulling power than any fighter in the world. And his new friends knew it. Even as he stumbled towards the exit, the Bomber could do it at the gate. He was part of everyone's life, part of the fabric of America. He'd grown up in the age of imagination, in the radio age. He was, to many, still perfect

– because they wanted him to be. They didn't want to think he was human, like them. They willed him to be superman. He beat Schmeling bad, didn't he? He beat everybody. OK, he got a little lucky against Walcott. Everyone's due a bad night. He's Joe, though. He'll be OK. Don't worry about Joe. Joe will be fine.

Except he had not been fine for quite some time, and would never be fine again. As the IBC embarked upon their sordid adventure with the semi-retired champ, as they started to drain boxing dry of its remaining credibility in these last days of proto-innocence, the game began to die around the edges, like a precious plant starved of water. Boxing would never be what it was, in the land of long-lost imagination. Nostalgia could not sustain it, could not beat off reality.

Whenever Joe thought about the good days, he thought about Chappie – that was the pet name they had for each other. If Chappie had been there, maybe he would have been OK. Joe said once, 'I guess I thought I'd be heavyweight champion forever. And I guess I thought Chappie would always be with me.'

Very soon, boxing was to be served a cold, unpalatable dish of reality.

Almost unnoticed now, Estes Kefauver slipped into the narrative, with a naive but firm conviction that there was an alternative to the dark side.

9

THE CAT IN THE COONSKIN HAT

In 1950 there were 150 million people proud to call themselves Americans. They were shiny-faced optimists most of them, who considered, with justification, that they were far better off, materially and historically, than the other 2.5 billion people on the planet.

Six out of ten Americans lived in cities. After the Depression, they'd left the small, James Stewart towns in droves and desperation. They'd stayed on in the nation's metropolises, unwitting but contented converts to a philosophy of urban regeneration. There was an epidemic of change. Trousers got tighter, morals a little looser. But innocence still held sway: men married, on average, when they were twenty-three, women twenty; two out of ten of those unions would end in divorce. (By millennium's end Americans were marrying later – at twenty-seven and twenty-five – but only six out of ten marriages survived. Older but no wiser, then.)

The further they got away from the war, the quicker the indiscriminate rush to the shops, and the products that moved most readily off the shelves reflected what Americans thought of as attainable, desirable and, according to the advertising diktats of the day, fashionable. When the homely, white-haired actor William Boyd popularised a rather naff TV cowboy hero called

Hopalong Cassidy, kids persuaded their parents to go out and buy Hoppy lunch boxes for them. There were hula hoops and yo-yos and bubblegum and the drink of the nation, Coca-Cola, which came with a trademark, a naming right that told the world it represented the United States of America and you could hardly go wrong by guzzling it freely. Fifty and more years later, kids would want Nanos and designer trainers. Their coke would come in lines not bottles. Diners Club became the card of choice and cachet in the fifties. Households lived on the never-never and, with inflation nothing more than that which filled a beach ball, there was no reason to think any of this would ever end.

Madison Avenue was the heartbeat of the advertising world and, by persuasiveness honed since the snake-oil days of the Wild West, its denizens determined where Americans would spend their dollars. In 1950, the suits of Madison Avenue – the Mad Men, as they called themselves – directed a third of all advertising money towards the pages of the country's newspapers. There was a paper in every town, big or small, and plenty in the big cities. The decade started with print still king. Twelve per cent of ads were on radio, a medium clinging on to fight off the new phenomenon, TV, which at that time had a mere 3 per cent of the pie. In all, the advertising industry spent $5.7 billion trying to seduce the nation in 1950.

This was unconstrained 'progress'. Bankers made the rules. The world, chiefly the United States, was embarking on an adventure in consumerism with no previous scale as a guide, where the only price worth noting was the price worth paying. Values moved from the beautiful to the big. Cars were long, loud and pink, with weird fins housing big red tail lights that looked like the eyes of a futuristic monster.

And standing tall in Washington at the start of this promising decade was Estes Kefauver, the clumsy loner from Tennessee, a Democratic senator who dreamed of becoming President of the

United States. The path he chose was righteousness. He believed in the basic goodness of his fellow Americans and reckoned they deserved to live in a country deracinated of poverty, prejudice and crime. One man among many he saw as a blight on his vision of this utopia was Frankie Carbo.

It is hard to imagine more strikingly dissimilar opponents in the boxing firmament than the senator and the mobster. Kefauver was awkward in more than one sense of the word, determined and blessed with boundless intellectual rigour and integrity but neither one of nature's born orators nor the silkiest of social beasts. Carbo was smooth, a shiftless charmer, a killer, liar and coward, a man irredeemably distant from the right path, or any suggestion of gaucheness that might arouse even incipient empathy.

Yet, what looked like a mismatch of which Tex Rickard would be proud turned out to be one of the fights of the fifties.

Kefauver made it his quest to lance the boil of organised crime that had poisoned American society since the days of Prohibition. He knew that the ghost-like figures at the heart of the Mob moved on waterfronts that traded in liquor, protection, debt collecting, gambling, prostitution and, increasingly, drugs. Boxing was just one of the Mob's red-light districts into which Kefauver peered. He made it his mission to hunt down Carbo and his cohorts, not so much because he cared about the boxing industry. He didn't. But it was an area of society he found riddled with rats, and the men who ran it the most rodent-like of all.

Kefauver was a decent man. He meant well. And at least he tried. He learned soon enough, however, that he was as human as the rest of us. In the end, this dedicated public servant was so seduced by the lure of higher office that he used the first inquiry as part of his bid for the highest office in the land. This was no crime in itself, and not that startling. But it skewed his normally astute vision. He made fundamental mistakes in his pursuit of the Mob, revealing an ego that few thought burned so brightly in him.

He risked going beyond his elected remit when he took it as a personal offence that such people could exist so blatantly outside the law. It was poor judgement by a trained lawyer, but wholly predictable in a politician.

Carey Estes Kefauver was born into comfortable but hardly ostentatious circumstances in Madisonville, Tennessee, in 1903, about a year before Paolo Corbo was born in New York City. Corbo would go through several names. Kefauver just dropped the Carey.

Estes Kefauver: it was a name that, through the flowering magic of television and against his instincts, would become known in every house in America. Kefauver was a gritty bookworm at school, more accomplished hovering over Shakespeare than when strapped into running spikes, or hurling the discus. He played in the football team at the University of Tennessee without creating a stir, and was glad to head for Yale and set about putting in place the building blocks of a legal career. He graduated, *cum laude*, in 1927 and practised law diligently in Chattanooga. He was, by all accounts, a fair and liberal man. His schoolmaster, H. L. Callahan, told his biographer, Charles L. Fontenay, that Kefauver treated poor black clients, for instance, 'with no assumption in voice or manner of superior wisdom and intelligence . . . always calm, kindly, courteous, genial and affable'.

Carbo, the devil incarnate, had already embarked on a life of murder and manipulation. He surely didn't have a prayer against such a saintly man.

Kefauver married, got into Democrat politics and, in 1938, full of New Deal zeal, was elected to the House of Representatives, then the Senate in 1949 and, in that chamber, he set about his mission: to hit hard and often at the growing phenomenon of organised crime. He was a senator for nearly fifteen years and was regarded warmly and with respect by his peers without inspiring wholehearted affection. Kefauver had a misleadingly stern countenance that turned back intimacy but it allowed him the space and time to pursue those political ambitions he reckoned were in his ambit.

Growing up during Prohibition and the Great Depression hardened Kefauver, made him even more focused than he might otherwise have been on killing the cancer of corruption he perceived had brought his country to its knees. He saw how brazen gangsters such as Lucky Luciano and Louis Lepke Buchalter were able to establish their secretive syndicate of crime bosses several states, untouched, unquestioned. It was the movement of contraband and ill-gotten money across state borders that made catching them so difficult.

Kefauver came to prominence through a remorseless application to challenge the torpor within the Democratic Party machine in Tennessee, a state ruled for fifteen years by Boss Ed Crump, a legal version of Carbo. Crump: it's a name Hollywood might have chosen – short, uncomplicated, unsentimental – for a smiling and insincere backwoodsman fixer from another era. And that is pretty much what the tyrannical Crump was, 'the old Red Snapper of Memphis', as contemporary accounts described him.

Crump had run the Tennessee Democratic machine for so long, Kefauver's challenge looked almost unpatriotic to the old guard. Crump picked on his challenger's personal idiosyncrasies. Abe Lincoln survived parodies of his homely ways, his lack of social grace. So too would Kefauver, particularly his naivety and lack of cool. Crump called Kefauver a 'pet coon for the Soviets' because he had once been photographed in a coonskin cap. Kefauver turned this childish, Cold War jibe back on Crump. He took to wearing a coonskin cap on a regular basis, not a bad move in a state where Davy Crockett was revered. Kefauver appealed to voters' sense of humour and their grown-up side. Eventually, he was rewarded.

This was the cauldron in which Kefauver refined the political skills that would serve him well when he made it to a position of real political influence. Voters took exception to the crudeness of Boss Crump's campaign on behalf of Senator Tom Edwards for the Democratic nomination for Tennessee. They didn't like the fact he tried to paint Kefauver as 'a Red', which was still a 'hanging offence' in the American psyche. They tired of the old man. They warmed to the young one, despite his naffness. And, at forty-four, the Abe Lincoln clone got his party's nomination for the Senate – and went on to make it all the way to Washington.

Politics in post-Roosevelt America provided young idealists with the moral platform to go after something better than the establishment had allowed to take root over many decades. America had come through boom-and-bust, the devastation of the dust-bowl thirties, reconstruction under a brave president – not to mention the demands of war, however removed the sound of guns and bombs were from their precious, shining shores.

There was a universal thirst for a better world from the moment the armistices were signed in 1945. Two international conflagrations were enough for any civilisation. The only way forward, from a left or right perspective, was through man's better nature. This was a notion widely held, nowhere more fervently than in Kefauver's breast.

After the war, a lot of young Americans did dream of something better than the arrangements and deals, the nods and winks of the old machine that had corrupted the system. Only a few would succeed. The bad guys didn't change, though. They still pilfered without censure. And that angered Kefauver, that rare political animal, a man determined not to waste his power. He did not intend to restrict his politicking to busting the Mob and their myriad rings, but that is the way it turned out in the end.

Once he got his hands on some serious levers of power, Kefauver worked with the fervour of the innocent. This was Mr Smith Goes to Washington. Kefauver was Jimmy Stewart without the good looks or the girl (but Lord knows he tried). He was straight as a five-bar gate, boringly diligent, scrupulous and hard-working. He would move from an undercard fighter to the main attraction in a surprisingly short space of time and, by 1950, found himself in charge of the Senate's crime investigation committee. This was not a fight even the Mob could fix . . .

The Thursday of 5 January 1950 was a quiet one in boxing for New Yorkers.

Over in Brooklyn, at the Broadway Arena, they rolled up to see Johnny Forte put away Hy Meltzer – a classic Wop–Yid face-off – in the sixth of eight. At the Sunnyside Garden in Long Island, the Italians were out in reasonable numbers to see their boy, Joe DiMartino, knock out Pat Kelly in the seventh. The Irish went home despondent.

But not everyone was talking about the fights that night. The following morning, down at Jacobs Beach, a few of the guys who read the front pages of the newspapers, got in a huddle to talk about this sonofabitch senator from Washington with the funny name. And in Washington, word was stirring that Senator Kefauver was messin' where he shouldn'ta been messin'. Kefauver had introduced a bill, giving notice to Carbo and the boys that he was going to not just investigate but destroy organised crime in the United States.

It had not been a straightforward decision for Kefauver to make. 'This was an explosive issue,' Halberstam writes, 'because the kind of crime Kefauver was going after had deep roots in every big city, and those cities were controlled by Democratic political machines.' They were the sort of machines favoured by the most powerful Democrat in the country, President Harry S. Truman. Kefauver was rolling the dice for high stakes indeed.

But he had the support of some influential liberal voices in the Democrat community, notably Phil Graham, who published the *Washington Post*. It was Graham's view that if someone in the

Down but not out: Joe Louis takes a tumble en route to a controversial win over Jersey Joe Walcott in their first meeting at Madison Square Garden in December 1947. 'Sorry,' the champ whispered to the loser later. The following June at Yankee Stadium, Louis stopped Walcott in the eleventh round of their rematch and then announced his retirement for the first time. It would be another three-and-a-half years before he finally quit boxing.

Uncle Mike: A month after Max Schmeling had knocked him out in the twelfth round at Yankee Stadium in June 1936, Joe Louis returned to the scene of his humiliation to take in the conciliatory words of the Twentieth Century Sporting Club's influential deal-maker, Mike Jacobs.

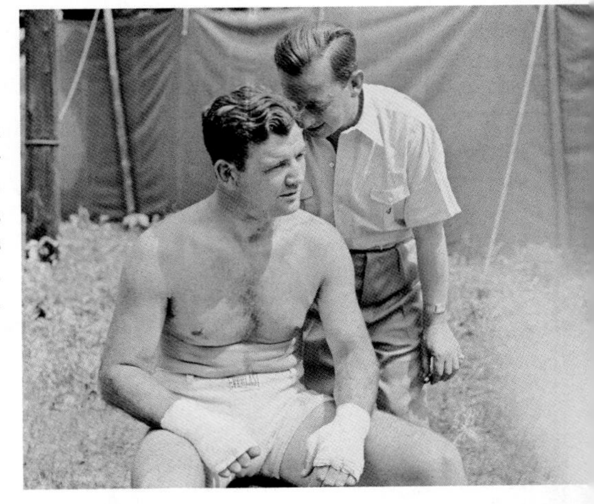

Uncle Joe: James J. Braddock would win the title from Max Baer at Madison Square Garden in 1935 and lose it almost exactly two years later to Joe Louis. Here he takes in the counsel of his manager, Joe Gould, at a training camp.

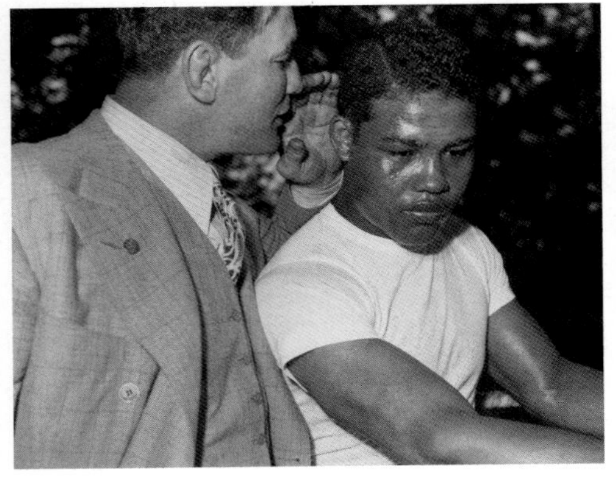

A quiet word: Braddock, comfortable in the knowledge that he and Gould were getting ten per cent of every purse Louis received for fighting for Jacobs, confides in Joe before his return fight with Schmeling in 1938.

Lucky Jim: Max Baer comes under pressure in his own corner in front of forty thousand incredulous fans at the Madison Square Garden Bowl on his way to losing the title to the lightly regarded James J. Braddock in June 1935.

Sweet revenge: Two years after losing to Max Schmeling, Joe Louis brings the German to his knees in round one at Yankee Stadium, with millions listening on radio around the world. The new champion looks on impassively from a neutral corner as handlers clamber into the ring.

A man of many words: Damon Runyon, pictured here at his typewriter in December 1937, wrote his way into the fabric of American culture through short stories that defined a fading era. He loved boxing – but his was an uncritical eye, as he chose to ignore the dubious activities of his many powerful friends in the fight game.

No Nonsense: Runyon called Dan Parker – seen here composing his thoughts for the New York *Daily Mirror* in 1953 – 'the most constantly brilliant of all sportswriters'. He was, though, not so tolerant of corruption as Runyon. No boxing writer did more to expose the Mob's links to boxing.

Tough Beach: This was the unprepossessing stretch of pavement opposite Madison Square Garden known as Jacobs Beach, originally the ticket office of the eponymous Uncle Mike. For two decades it was the meeting place of every major face in the game.

Beyond the Shor: When business was done at Jacobs Beach, or in the Garden, boxing's heavyweight decision makers and their celebrity friends repaired to Toots Shor's. Here, in December 1954, Toots looks on as the actor and comedian Jackie Gleason roars drunkenly into song.

Not so smart: This was the fight that helped bring the Mob to its knees – Jake La Motta, on the right, taking an arranged hiding from Mobbed-up Billy Blackjack Fox at Madison Square Garden in November 1947. The Raging Bull would not admit the fix until he went before a Senate inquiry in 1960.

Crumbling Rock: Rocky Castellani, on the left, and Ernie Durando, brought Madison Garden to life one January night in 1950 – but not as they intended. When the referee Ray Miller gave the fight to Durando, Rocky's Mob-connected manager Tommy Eboli stormed into the ring and assaulted him.

The Cuban Hawk: Kid Gavilan outpoints clean-as-a-whistle New Yorker Billy Graham at the Garden, in November 1950, making it one-all between them. Their third meeting, the following August at the same venue, saw Gavilan win one of the most controversial decisions of the era.

Sweet success: Johnny 'Honeyboy' Bratton levels New York left-hooker Joe Miceli at the Garden in October 1952. Bratton signed with mobbed-up New York manager Hymie 'The Mink' Wallman to move his career.

Garden II: Built on the site of the original Garden, at 26th and Madison Avenue, this was the creation of Stanford White, brilliantly over the top, and housed boxing and other attractions until it was torn down in 1925.

Garden III: This was the 'old garden' of legend, on Eighth Avenue between 49th and 50th Streets. The gathering here is not for boxing, but a meeting of the Young Communist League in January, 1939, quite possibly attended by Budd Schulberg, who wrote *On The Waterfront*, a poetic testimony to the power of the mob over boxing and ordinary men.

Garden IV: The current site, on 33rd Street between Seventh and Eighth Avenues, trades on the history of the Garden, but is not much loved by those who remember Garden III – or the old Pennsylvania Station that made way for it.

Democratic Party did not take on the mobsters, the Republicans would. He urged Kefauver to lead the charge and, with the presidential election two years away, he tempted him with the sort of words any ambitious politician in America at the time would want to hear: 'Don't you want to be vice president?'

So 'The Keef' set off on an odyssey that was to take him to the very steps of power and, eventually and almost by accident, would eat away at the heart of sport's worst-kept secret, professional boxing.

His strategy was double-edged. In pursuit of the evidence he needed, Kefauver took his investigating team to fourteen cities across America in ninety-two days. This not only disturbed the local political movers and made the gangsters nervous, it earned him acres of space in the prints as he built his political profile. He might have come across sometimes as a hillbilly senator, but he was shrewd too, a Yale graduate from a privileged Tennessee 'old money' family. Kefauver was on his way.

Kefauver's opponents were many, and largely invisible. The storm troopers of the Syndicate were those bloodless killers in Murder Inc. – which functioned mainly between 1930 and 1940, and included Carbo, Bugsy Siegel, the feared Albert Anastasia and scores more faceless thugs in sometimes competing factions who would otherwise have been on the dole with their 15 million compatriots. Buchalter went to the chair and Siegel went in a hit; Kid 'Twist' Reles, who, for two years, squealed like a pig stuck forever in a swinging gate to send key capos to the chair, had the misfortune to fall six storeys to his death from a room with several police guards at the door.

One of the cops assigned to protect the canary Reles was Captain Frank Bals – a curious choice, given Bals was a known bagman for gettable city politicians. It emerged years later about $250,000 a week passed through his hands as hush money. Bals was also close to Bill O'Dwyer, whose public face was as a campaigning district attorney on the trail of the Mob. Bill didn't have much luck catching any bad guys. And, after he became Mayor of New York, he made Bals Deputy Police Commissioner. A year into his second term, O'Dwyer was forced from office when it was revealed he had taken orders on public appointments from one Frank Costello, a mobster with eyes cold as a fish's – and Carbo's boss.

On balance then, you'd have to say Kid Twist was unlucky to end up in that particular room.

The killing might all have been in-house, but it was no less chilling for that. Kefauver, the so-so college athlete, the 6ft 2in geek, saw it all from a distance in the thirties but, come his time, he was to be the worst nightmare these hard men could imagine. And his time was the fifties.

As his caravan of inquisition got closer to New York in 1951, the nation woke up to the realisation that the senator was serious. They were about to see it all for themselves, in the comfort of their own living rooms.

On the morning of Monday 12 March 1951, Kefauver picked through the *New York Times* over breakfast. He glanced at the sports results. Then he got to wondering how the newspaper guys might regard his performance later that day on television. He had only been told at the last minute the hearings were to be broadcast and he wasn't totally prepared for it. What would the cameras do to his plain features, his un-Hollywood presence? Would he make a total fool of himself in front of millions of people?

The medium had pussyfooted around inquiries like this in the past – most famously the House Un-American Activities Commission hearing on the spy Alger Hiss. TV was still an infant, and that was mainly on local outlets anyway. Later that Monday, Kefauver was going to become a national television celebrity, seen across the nation as he went after the hoods.

He was nervous. Never comfortable in front of a microphone, he decided he would keep his pronouncements brief and dry, like any lawyer would, trying to pin down and confuse the accused with unanswerable legal logic. He would not showboat. But although Kefauver had initially been sceptical about appearing for so many hours a day on television, his ambitions got the better of him. All the time he had the words of Phil Graham ringing in his ears: 'Don't you want to be vice president?'

He did. Desperately. This was his chance. The boy from Chattanooga was getting closer to the White House. And even if he didn't make it, he would leave his mark on American politics.

That first broadcast went out to twenty cities, mainly in the East, but also reaching into the fabled Midwest, the heartland of America, where political reputations were made or buried. The audience was small at first, especially in the mornings, when only 1.5 per cent of American homes had their sets turned on. But word spread. The phenomenon of watching gangsters sweat on TV was too much for people to resist.

Time magazine got in early, sponsoring the hearings in New York City and Washington. As Halberstam said, Kefauver had lucked out in his timing: any earlier 'and there would have been no audience; a few years later, there might have been less excitement, for people might have been more blasé'.

They were still wide-eyed and innocent that Monday in March of 1951. The following day, the guys at Jacobs Beach gathered again for coffee. They were full of the usual stuff about the fights the night before: Paul Pender should have done better than a draw against Joe Rindone up in Boston, and what about Tony DiMarco – how did he get knocked out on the undercard by that bum Chic Boucher? Over at Laurel Gardens in Jersey, Bobby Fenty got past Johnny Kamber in the main event, and up in Rhode Island, Roland LaStarza was still doing good. He decisioned Keene Simmons over ten. 'He'll get a shot at the title. Don't worry about that.' Sure he would. LaStarza's people were well placed. So were those of the man he would one day challenge, Rocky Marciano.

But what was this malarkey on the front page? 'This Kefauver guy,' one said, 'he's really breaking balls, ain't he? I don't like that. At all.' They sipped their coffee, nodded, and didn't say much more. Then: 'Yeah, and do ya know it was on the TV?' said another. 'Right there on the fuckin' television. Hey, and you know who's in the dock today? The man. The man himself.'

Later that day, dressed better than anyone in New York, Francisco Cataglia walked into the third-floor room of the legal chambers at Foley Square and a hush fell over proceedings. Known to his friends and enemies as Frank Costello, America's most intriguing criminal had been cornered at last, for all of the nation to see – or at least all of America with television sets. And Costello, the most feared man in town, Mr Cool, the gangster whose word was law among his peers, was even more nervous than Kefauver. There followed an absorbing piece of theatre, a strategy that was to backfire spectacularly on Costello.

Alongside Kefauver and his assistant, Rudolph Halley, sat the other members of the investigating team, Herbert O'Conor from Maryland, Wyoming's Lester Hunt and Charles W. Tobey, a New Hampshire senator with a particularly sharp tongue.

Suddenly, Costello felt intimidated. He was not in control. And, looking at the cameras, he did not think having his face on television was such a great idea. He did not seek this out; they had subpoenaed him. Frank did not want to come across as a common hoodlum on national television when he was trying to do mainstream business outside the courthouse.

His lawyer, George Wolf, delivered one of the most memorable pleas every recorded on American judicial log: 'Mr Costello doesn't care to submit himself as a spectacle.' The packed gallery struggled not to laugh out loud. Cops nudged each other. Lawyers tapped the desk with their pencils. But the committee agreed. They could not afford to lose their star witness. They had no idea how their decision would play so dramatically in their favour.

How the guys at Jacobs Beach and other Mob hangouts enjoyed that one. Frankie didn't want make a spectacle of himself. As ever, he wanted someone else to do his dirty work. He wanted to fade into the background, keep pulling his strings. TV was not for Frank Costello.

So, as seven out of ten New York TV sets were tuned into the hearing – double the ratings of the World Series the year before – Costello sweated like a rapist for the inquiring lenses of the new medium. Except, the court having acquiesced to his lawyer's plea, the cameras did not concentrate on his grey, pallid features at all. They did better than that: they went down to the desk on which he rested his hairy, nervous hands and watched him squirm digitally. His fingers twitched. He drummed the table. Spare paper was crumbled and torn. He was bricking it by hand.

Word spread. This was serious stuff with serious consequences. Costello, the criminal mastermind, was turning to mush under the polite interrogation of Estes Kefauver and the more robust questioning of Senator Tobey.

The panel wanted to know about movements of money across state borders, about unaccounted-for cash, how Costello earned a living with no visible means of support, how he seemed to influence the workings of racetracks and other sporting events. They reached way back into his past, which might not have been the smartest move as it gave him the opportunity to repeat all day, 'I don't remember.' It had always been the Mob's fallback defence. Amnesia was contagious in the underworld.

Nevertheless, it was compelling drama. 'Never before', *Life* magazine commented, 'had the attention of the nation been so completely riveted on a single matter. The Senate investigation into interstate crime was almost the sole subject of national conversation.' And Kefauver was the first reality-TV star.

In the end, they got Costello on some fiddling perjury counts. He'd lied about bootlegging. He'd failed to tell anyone he once went by the name Frank Severio. He'd lied about having a private investigator sweep his premises for wiretapping.

The following day, Costello was a wreck. He said he was ill, didn't want to talk any more. He said to Kefauver: 'I want to testify truthfully and my mind don't function . . . With all due respect for the senators – and I have an awful lot of respect for them – I am not going to answer another question . . . I am going to walk out.'

He was threatened with contempt and came back the next day, sniffling. He left again. In the end, it was all a bit inconclusive. But it opened up crime to the country. America peered inside the box and saw, for the first time, in living black and white, the workings of the Mob and Tammany Hall.

Once he got into his stride, Kefauver milked his celebrity shamelessly. He wrote a book, *Crime in America*, which the *New York Times* listed among its best-sellers for three months; he was on magazine covers across the country; he went on TV as a guest on a quiz show; he was offered a part in a Humphrey Bogart movie called *The Enforcer*. The Geek was on fire.

Later in 1951, he wrote an oddly detached and lively account of the last eight days of his investigation and his impressions of Costello. The language is not that of the dry government report; but then Kefauver did have his eyes on higher office.

He revelled in what he perceived, with some justification, was the humiliation of Costello and that of Bill O'Dwyer, the Irish immigrant who'd risen from patrolman to become Mayor of New York.

'To me,' Kefauver wrote, 'the position of Mr O'Dwyer was particularly lamentable – a melancholy essay on political morality.' He produced testimony that proved the ex-mayor was taking $10,000 kickbacks from organised labour.

But there was only one star. 'Of all the witnesses from the crime world summoned before us in New York,' the senator intoned, 'Frank Costello was the focal point of interest. The remarkable thing about his appearance before us is that Costello, the vaunted "prime minister" of the underworld, made himself out – and needlessly so – to be unintelligent.'

This seemed a crude misreading of street smarts. Costello did not care about his image in the dock; he just wanted to get the hell out of there.

'He also exposed himself as a whiner and something of a cry-baby,' Kefauver wrote. 'He did not even display particular courage for, after threatening twice to walk out on the committee, he lost his nerve, came back and abjectly answered most of our questions in a manner that indicated his defiance was gone. I do not pretend to understand how the minds of mobsters work, but I should think that Frank Costello, by his senseless performance before us, would forfeit the respect of even the hoodlums to whom he has given orders.'

Kefauver was right. Costello was, on the face of it, America's most famous gangster since Al Capone. And he was popular among his own kind – murderers, extortionists and pimps who dressed themselves up in the quasi-respectability of belonging to New York's Luciano family.

His friends in the fight game were no better.

Costello, for much of his adult life, was at or near the apex of this criminal pyramid, at least on the East Coast. And his willing lieutenant was Carbo, mysterious, sometimes urbane, more often cold and heartless. He was a man who had the intelligence to avoid a life of crime but chose not to. He had no excuses. Costello could not have wished for a better pal, although he would soon tire of the power struggle at the top of the pile, and retired in favour of Gabe Genovese. Carbo, however, retained his criminal thirst.

Carbo, or Corbo as his father Angelo was called, might have been just another son of immigrant Italians who would go on to make the most of his gifts from an unpromising start. That he did not do so by the legitimate route was no reflection on his parents,

Angelo, a labourer riddled with TB, who would die at fifty-four in 1931, and his mother, Clementinia Petrono. She was the classic Italian momma; little else is known about her, except that Paolo, or Frankie as he was more universally called by then, looked after her until she died in 1942. Not that he deserved a medal for that.

It seems there was nothing Angelo or Clem could do to curb the tearaway instincts of their smart, cruel son growing up on Second Avenue. He thieved and he bullied and hung out with already corrupted older boys who earned their living the easy way. At eleven, young Frankie was sent to reform school. When he came out, he chose the easy way too.

He was not big, but he made much of his squat physical presence, five foot eight and just over twelve stone. His hair was already greying, which may or may not have been the reason one of his several nicknames was Mr Gray. More probably it was a combination of that and his ability to fade into the background, contactable only when he wanted to be. He was a sinister presence from an early age in his career as a criminal, and traded on the mystery of his persona, self-conscious and self-confident at the same time.

New York police were on his case from the off. As Jim Brady observes in *Boxing Confidential*, nobody is sure how many times Frankie was arrested as a punk hoodlum on the way up, because the records went missing after his notoriety superseded his anonymity. Nevertheless, the FBI eventually compiled a quite staggering half a million pages on the man who would become Frankie Carbo.

Part of the problem, of course, was that Mr Grey was also known at various times, for obvious reasons, as Paul L. Carbo, P. J. Carbo, Frank Fortunati, Frank Marlow and, more colloquially and almost certainly not to his face, 'Jimmy the Wop'. This was a man in search of an identity, but who really wanted none. He was a complex product of his anarchic times, never absorbed into mainstream society but all the while giving the appearance of being totally embedded as a respectable citizen. This is at the root of the ambivalence about Carbo. He had that peculiarly elusive street skill of making all who came into contact with him (or nearly did) believe he was benign, charming, on their side.

So overpowering was his 'grey' personality, people who never met him but mixed with his henchmen convinced themselves they knew him better than they possible could have. They accepted the general judgement that 'Frankie Carbo was a very nice guy'.

You could not entirely blame them. They lived within a frame of moral reference that was different to that of most people. Some were there, in the boxing business, by choice – but most were not. They were flung into the demi-world of screwed values and soul-sapping compromise by circumstance.

All the while, Frankie Carbo was pulling the strings that made hundreds of boxers, managers and promoters dance, although they didn't always know it. Carbo had been doing this for a long time, probably as far back as the early thirties.

A delinquent of long standing, Carbo came more regularly to the notice of the New York police as he stretched himself for Costello and others. In his circle, his CV was impressive.

In 1928, now a hardened twenty-four-year-old pro who'd long since stopped listening to his parents, he tried to put the frighteners on a cab driver from whom he had demanded protection money. The cabby wasn't having it and, with a muscular friend called Albert Webber, confronted young Carbo. Webber was shot dead, and Carbo was charged with his murder. This was reduced to manslaughter, for no apparent reason, and Carbo was out of Sing Sing after serving just eighteen months of a four-year sentence. It was to prove a rare victory for the police, if a rather thin one.

Carbo's career as a killer was secured. His reputation was in place. There was no turning back. In 1931 he escaped prosecution for the murder of a New Jersey liquor trader called Michael J.

Duffy. Carbo's alibi was simple – and hilarious: he was no hitman, he told them, he was a $50-a-week bus inspector and, at the time Duffy was gunned down in Atlantic City, he was, according to FBI records, 'undergoing treatment at an ear clinic'. Hear no evil, see no evil.

Carbo had been drawn deeply into Murder Inc. as a trusted operative. At the head of this self-important but highly dangerous killing gang was Louis Lepke Buchalter. Around Buchalter in various parts of New York were Charles 'Lucky' Luciano, Vito Genovese (the cousin of Gabe), Abner 'Longy' Zwillman, Meyer Lansky and, the deadliest and scariest of them all, Costello.

These were Carbo's associates. Their day job was extorting illegal protection payments from innocent citizens, blackmail and generally organising a climate of illegality throughout the reaches of the established law enforcement bodies, as well as in the darker corners of municipal, state and federal government. By night they sold hooch, gambled and fixed fights – which is where Carbo always felt most comfortable. He really loved boxing – even as he was orchestrating its utter corruption.

The murdering continued. Carbo is said to have taken part in at least half a dozen in his lifetime. Nobody is sure, because he covered his movements so deftly, and had the silence of timid underlings as a shield.

Carbo got into the fight game in a major way about 1933. He did it by controlling one man: Mike Jacobs. Acting largely on the instructions of Costello, who also loved the fights, Carbo muscled in on the operations of Uncle Mike with sledgehammer subtlety. Perceiving no resistance, he went on to formalise the relationship. The Garden was, without doubt, at the centre of the boxing world and any manager or fighter who wanted to do business there had to play ball with the likes of Carbo, or his associates, of whom there were quite a few, most notably Owney Madden. There was very little Jacobs could do about it.

Some years later, in 1939, Harry Markson, the Garden's loud PR man, declared that Jacobs stood for everything straight and fair in the boxing business. 'Mike Jacobs has made a genuine effort to reduce to zero the influence of gamblers in boxing and has taken strong measures to prevent them from operating at the Garden . . . Boxing today is immensely more honest than it was fifteen or twenty years ago.'

Nobody else thought so. Especially those managers who kept their eyes open and their mouths shut. They knew that the control was centralised because, if they didn't agree to sell a slice of their fighter to a Mob-favoured partner, they didn't get to fight at boxing's most famous hall, or one of the associated offshoots in Philadelphia, Detroit or Chicago, the only places a good fighter could make decent money.

In a way, it was fitting that Kefauver was not the man to land the finishing punch in the later rounds of his fight with Carbo and his followers. That honour belonged to the FBI, an organisation closer to the business end of crime. But Kefauver will have reckoned he contributed significantly to nailing the pompous psychotic who ran boxing from the shadows. Unlike some of the contests Carbo arranged in a dozen or so years as a malign influence at Madison Square Garden and other boxing halls of ill repute, the result was no foregone conclusion. It took every drop of Kefauver's resolve, and the cooperation of the gods, to hound Frankie.

Kefauver, I was reminded by the New York writer Thomas Hauser, is pronounced on that side of the Atlantic with the emphasis on the first syllable, the second two floating fluffily away almost unnoticed – which makes his name sound a little like a muffled half-sneeze if said quickly. And more than one bad man caught a cold from Estes Kefauver, mostly without knowing they'd been infected. He was a curiously soft presence in any company. In fact, not much about the appearance or mien of this dogged crusader hinted at heroism.

Jim Bishop wrote in the *New York Journal-American* in 1956, the year Kefauver ran for president, that he 'looked like an overly alert horse that was getting Dexedrine in his oats'. Certainly, the senator did not lack for energy in the race, however ungainly his gait. It was Carbo's misfortune that they were in office at the same time.

Kefauver's blows were not always accurate; indeed, the conclusions of his first inquiry into organised crime, delivered in 1951, were naive and, in some instances, laughably wide of the mark. A dogged crusader, he would try again ten years later.

At the end of round one, the Keef and Carbo were level on points. However, the better cards were in the gangster's hand – thanks mainly to Joe Louis.

10

THE RISE OF THE ROCK

Once there were two fighters, American welterweights, improbably called Fast Black and Jabbing Jake. No surnames. No antecedents. Pretty much nothing is known about them now, where they came from, where they finished up. Indeed, not a lot was known about them the night of 14 March 1944, when they met at Dorsey Park in Miami, Florida. Jake stopped his man in four.

They were spectacularly unsuccessful prizefighters. Jake had three paid contests in his life; the victory over Black was his one success. Black fought just seven times, winning twice – but one of those wins was over a very good fighter called George Monroe Costner.

Sugar Costner, as fight fans in his native Cincinnati and myriad venues throughout the country knew the handsome six-footer, was in a different league altogether. Yet somehow he lost to Black – on points over six rounds at the Coliseum Arena in New Orleans in 1941 – before going on to box four of the finest fighters of his or any other era: Sugar Ray Robinson, Jake La Motta, Kid Gavilan and Ike Williams. He lost to Robinson and La Motta when he was near his peak yet, remarkably, beat Gavilan and Williams at the end of his career, in the early months of 1950.

So, how is it, then, that Fast Black, who lost to a fighter even more anonymous than himself in Jabbing Jake, could beat a fighter who beat Gavilan and Williams?

The charitable view is that fighters were hungrier then, that unbeaten records meant little. Good fighters lost fights – simple as that. Conspiracy theorists and cynics, meanwhile, reckon you just couldn't trust any result. While not wishing to impugn the character or demean the achievements of Mr Costner, certainly there were more than a few fights in Gavilan's career that had a malodorous whiff about them. And Williams, to his cost, went on the record many times to suggest he'd been stitched up by gangsters and gamblers.

As the fifties got under way, Costner found himself top of the bill at Madison Square Garden. It was a Friday 6 January, the venue's first promotion of the year, and Sugar Costner was afforded the benefit of the doubt in a majority decision over Gene Burton, whose log stood at a respectable forty-nine wins, nine draws, nine losses. Within a month, Burton, who'd been good enough to draw with Gavilan two years previously, and outpoint Williams the year before that, reversed the decision over Costner.

You might have thought from these permutations that such variable results suggested all was on the level in boxing, that one decent fighter had as good a chance as another in just about any contest. What it also points to is a lack of any consistency in form – and that is where bookmakers always make money.

Professional boxing in the fifties was a lottery. Compare results with today, even (not exactly a golden era), and you will find upsets then were far more commonplace. This is what confounds historians.

It was getting to the point where fight fans couldn't trust what their eyes could see. They pretty much knew that La Motta had gone in the tank against Fox, that Graziano was running scared of the Mob and that good fighters such as Billy Graham weren't getting the breaks because they didn't have the right connections.

A sense of helplessness had enveloped the sport. It was worse than the twenties and thirties, worse than the forties. What the Twentieth Century Sporting Club had established at Madison Square Garden since the rise of Mike Jacobs was a loose, unspoken control. Every fighter wanted a title shot in the Garden but, in the early days, if Uncle Mike didn't want to play, there were other places to go, other promoters to talk to. Gradually, however, over the years those options dried up. Mysteriously, managers who crossed the guild in New York found doors shut in Philadelphia. That was Blinky's patch. A fighter who lost his licence in New York, such as Graziano, merely moved out of town for a while and, when he returned, all was forgotten. There seemed little point in hounding the Mob, because their tentacles had spread so far. This was not just a New York problem any more. It was a nationwide conspiracy to defraud, spawned in the Garden and spread from Philly to Chicago to LA, and all points in between.

This was the curse that Kefauver feared most in the underworld, the virus-like spread of organised crime. For years the FBI's all-powerful chief, J. Edgar Hoover, had said there was no such thing. He did not want to recognise the rise of the star thug and his army of undetectable foot soldiers, even though he knew who and where they were. It suited Hoover to deal with the Mafia as if it were just another outfit of miscreants, rather than the out-of-control monster it had become.

In such a fetid atmosphere, with Kefauver initially drawing a blank against the mobsters, professional boxing just went about its business. Now, with Jacobs Beach populated by the same gangsters but under a new regime, the Garden was in the hands of Norris and the IBC, with Carbo and Palermo exerting their bullying influence from offstage, and Gibson lending a patina of respectability to the operation with his legal and cultural heft. It is still hard to believe, having met him, that Gibson got caught up with these shysters, but he did. And he very nearly went to prison with Carbo and Palermo.

Most obvious among the great disempowered was the former heavyweight champion of the world.

Ten fights and nearly three years after he'd announced he was quitting the ring, Louis met Marciano in the Garden. Joe had not been champion since surrendering the title to Ezzard Charles at Yankee Stadium a little over a year earlier. He had not been a convincing imitation of himself since 1941, when he came from behind to knock out Billy Conn the first time.

Now, balding, soft and confused in his private life, Joe was in against a man whose ring name could hardly have been more apt. Rocky was hard, from chin to fist, with a will to match, and a squat, rolling and bobbing technique that had brought him thirty-seven wins from as many contests since his professional debut in his home town of Brockton, Massachusetts, only five and a half years earlier. He'd needed the officials' judgement just four times, although he came desperately close to losing to Roland LaStarza in the Garden in March of 1950. Not everyone was convinced about the credibility of that result. One New York fight writer called the decision 'paper thin and exceedingly odd'. The scores, at first inspection, had it a draw, with the referee scoring it five rounds apiece, Arthur Schwartz giving it to Rocky 5–4 with one even, and Artie Aidala likewise, for LaStarza. However, under New York rules, the referee went back to his scorecard and made it 9–6 on points for Marciano under the supplemental scoring system.

It was not one of Marciano's better nights and the crowd made him aware of their disquiet. 'Fix!' was the ritual cry that filled the room. There was no proof this was so, but more than a few old sages wondered if the unsubtle hand of Carbo had intervened – without Marciano's knowledge, of course, according to his friends.

Lou Duva, one of Rocky's closest associates, won't hear of it. For him, and every Italian American even vaguely interested in boxing or in the magic of Marciano, Rocky could do no wrong.

Yet Carbo was always on the scene when Marciano was about. It was the same with La Motta. The same with Frank Sinatra, in fact. Unfairly or not, anyone in the fight game with even a tangential Italian connection attracted the interest not only of the Mob but of everyone else looking at the Mob.

When I spoke with Duva for this book, in 2004, I told him about a conversation I had a little earlier with a guy called Rollie Hackmere, a good amateur and professional flyweight from the fifties who later trained heavyweight Doug Jones and Bob Foster, the world light-heavyweight champion.

Rollie, an upright and dapper gentleman in his seventies, called me aside one night at a meeting of Ring 8, the New York ex-boxers' association, and said, 'I knew Rocky. He was some guy. Do you know he fought his brother, Sonny, nine times? No. I bet you didn't. How about that?'

'Well, how about that?' I asked Lou, who kept in regular touch with Rocky until he died in a plane crash in 1969.

Duva, not a man worried about shouting 'bullshit!' very loudly right in your face, didn't miss a beat. 'Yeah he did, yeah. He fought Sonny. They was fights in the smokers. So what?'

So what, indeed. That was the way things were, apparently. Smokers, informal bouts organised in small clubs, usually featured amateurs and winners received either small cash prizes or trophies. It was considered a good learning experience, away from the scrutiny of the media, where boxers could learn the tricks of the trade. But beating up on your brother was, to say the least, an unconventional way to do it.

The night he looked across the ring in the Garden at Louis, Rocky might have reflected on those smoke-filled nights and how far he'd come. There was no title at stake, just the chance, the obligation, even, to retire a legend. And, with all the clinical detachment he might have brought to punching brother Sonny in the nose back in Massachusetts, he set about destroying what was left of Joe Louis on Friday 26 October 1951. Anyone who was ringside would have had to work hard not to cry when Marciano delivered the last of many rights he crashed on to Joe's old head in the eighth round. Louis was virtually out when the goodnight punch landed, having shipped an equally concussive left seconds before. Hands down, head lolling, he took the shot and his noble frame flew through the ropes, landing on the apron. He was slain, eyes dead, spirit broken, as the lights above bathed his worn skin. He did not move until assisted by the officials whose pens and notebooks he had disturbed. It was to be his farewell position in a sport he had graced with power, skill and dignity since the night in Bacon's Arena, Chicago, in 1934 when he dropped Jack Kracken with a left hook. Then Joe did to Jack what Rocky had just done to him, and put him through the ropes on to the lap of the local commissioner. It was Jack's last fight too. Joe knew all the emotions of the fight game, from top to bottom, from start to finish. He identified as much with Jack Kracken, an obscure Norwegian whose real name was Emile Ecklund, as he did with Rocky, whose real name was Rocco Francis Marchegiano.

But Joe's exit was one fans could at least accept as real. Their eyes did not deceive them this time. It was shudderingly final. There would be no more comebacks for Louis, not in the boxing ring anyway. The lights went out on an era when Rocky knocked Joe out of boxing, literally.

Marciano's reign would be distinguished, yet it never captured the imagination as did that of the Brown Bomber. The heavyweight division in the fifties would be marked by Rocky's indomitable determination and disregard for niceties, his departure, unbeaten, leaving another chasm for lesser men to fill. With the connivance of the IBC, they would do just that.

Interest shifted to the middleweights, where La Motta ruled. He had the title now, reward for faking it against Fox in '47, but it was a troubled and curious reign from the very beginning. After beating a one-armed Cerdan to take the belt in June of '49, he then fought Cerdan's compatriot Robert Villemain over ten rounds in a non-title bout in the Garden. This was not unheard of, but it was unusual. What was also unusual was that La Motta, at the height of his powers, lost emphatically to the Frenchman – about as clearly as Villemain had trounced La Motta earlier in the year without getting the verdict, also in the Garden. Such was the furore over their first meeting that the New York State Athletic Commission chairman Eddie Eagan shocked the boxing fraternity when, three days after the bout, he suspended the referee Harry Ebbets and one of the two judges, Harold Barnes, 'indefinitely'.

It was Mr Eagan's opinion that 'in view of the international importance of the bout and the fact that their cards were contrary to the viewpoint of practically all those who witnessed the contest, especially the members of the commission, it was felt that disciplinary action was required'.

Ebbets scored the rounds 6–4–2 for La Motta, Barnes gave it to him 6–5–1, while Charley Shortell saw it 7–4–1 for Villemain. Twelve of thirteen fight writers polled at ringside gave it to Villemain.

The Garden crowd, their intelligence serially abused now, again bellowed 'Fix!' into the New York night. The talk at Toots Shor's later was similarly sceptical. There was no proof that La Motta had been involved in his second fix in two years. Nor in the return with Villemain, when the champion looked seriously underdone. The referee Ruby Goldstein gave it to Villemain 5–3–2, while Young Otto and Joe Agnello both saw it 7–3 for him. There were no arguments from La Motta or his handlers – but a lot of crowing from the Frenchman's corner. He would never get his hands on the title, nor even have the opportunity to do so (although his points loss over fifteen rounds to Robinson the following year in Philadelphia was for something called the Pennsylvania State world middleweight title). But he felt vindicated at last in beating La Motta, and he had an unusually passionate Associated Press writer trumpeting his case the next day in a report that departed from the dry style that normally constricts wire-service writing:

> Robert Villemain officially has evened his score with Jake La Motta. Now the scrappy little Frenchman wants the middleweight champion to put his title on the line. Villemain, a 5ft 6-and-a-half-inch bundle of perpetual motion, poured it on the slowed-up Bronx Bull last night to win a unanimous 10-round decision in MSG. That squared accounts for the 25-year-old Parisian, who licked Jake in the Garden last March but lost the decision. The verdict, one of the rankest in years, resulted in the suspension of the two officials who voted La Motta.

Indeed it did. But not 'indefinitely', as Mr Eagan ruled. Ebbets, who'd boxed in the twenties and briefly managed a football team at Rockaway Beach, New York, where he was born, continued as referee and judge for many years.

Barnes, too, returned. But, in 1958, he surprised many people when he told the *Saturday Evening Post* boxing had 'fallen into moral turpitude and physical derogation'. Barnes, who'd been an official since 1921 and refereed more than a hundred world title fights, many of them at the Garden, resigned. He was an optometrist.

It didn't make the *New York Times* – or even the *Tecumseh Countywide News* and *Shawnee Sun* in Jess Willard's home town of Pottawatomie, probably – but, on Wednesday 29 August 1951, Neil Armstrong, who would go on to take one great leap for mankind, flew his first mission as a pilot in the US Navy. Six months after learning to fly.

He rode shotgun for a photo reconnaisance plane over Songjin as the Korean War struggled into the American consciousness. America didn't want to read about Korea. It was a war tacked on to a war. It was far away.

'Unlike Vietnam in the next decade,' writes David Halberstam, 'it did not come back to America live and in color on television. The nation was not yet wired and Korea, so distant, the names of its towns so alien, did not lend itself to radio coverage, as did the great war that had preceded it.'

General Matthew Ridgway, writing forty years later, called it the Forgotten War in his book about a conflict nobody really understood. It was a good rehearsal for Vietnam.

No, America wasn't that interested in young Armstrong's flying debut for his country, or the war in which he and thousands of perhaps similarly nonplussed servicemen were getting involved. And killed. It was summer and the sun was shining. Girls in starched skirts and white socks were listening to Perry Como and the Fontane Sisters singing, 'It's Beginning to Look Like Chistmas'. In Korea, it was beginning to look like they would see more than one Christmas with their young heads stuck in a trench.

Back home, Americans had their heads stuck in Mickey Spillane's novels, and occasionally they'd read the paranoia about communism that Senator Joe McCarthy was starting to peddle through eager journalists. The country was awash with fiction. As George Reedy, the veteran United Press reporter, remarked about McCarthy: 'Joe couldn't find a Communist in Red Square. He didn't know Karl Marx from Groucho.'

The country wanted to have fun. And Kid Gavilan, the Cuban Hawk, was up for it – if you were sitting on the right side of the ropes, or were a good-looking girl in the Cotton Club. 'The Keed' threw his famous bolo punch – supposedly honed in the cane fields back in Berrocal, Camaguey, but probably born in the imagination of his publicists – with such panache, the ladies swooned and the guys put down more bets on a Gavilan win.

That Wednesday night in Madison Square Garden, as Armstrong steered his jet into the Korean skies, New Yorkers could take their mind off weighty matters on the other side of the Pacific and nearer home. They rolled up with loud enthusiasm to see if Gavilan could hang on to his world welterweight title against the most local of local boys, their own straight-as-a-die Irishman Billy Graham.

They'd already met twice, the Cuban and the kid from the East Side, Gavilan winning each time. There wasn't a lot between them in boxing ability, but the Cuban had more clout at the gate. He was charismatic, a lothario, a nightclubber, a man not scared to live life in a decade when life had been handed back to people for just that, a star who lit up the pages of the New York tabloids. He was exotic. Graham was solid, decent, a consummate, professional boxer with good skills and an admirable work ethic.

Promoters took $34,419 from 8,137 punters. And this is how James P. Dawson of the *New York Times* saw it the next morning . . .

> The writer scored the battle nine rounds to six, for Gavilan. The champion was entitled to the first four, eighth, ninth, tenth and eleventh and the thirteenth. Graham won the fifth, sixth and seventh rounds, the twelfth and, in a closing burst of brilliant boxing, the fourteenth and fifteenth.
>
> Graham's superior boxing skill carried him through the rounds he won. In these sessions he outboxed the champion . . .
>
> In the twelfth Graham's punches were cleaner and sharper and he grazed the jaw several times with rights. Through the fourteenth and fifteenth rounds Graham, throwing all caution to the winds, reached the champion's

> jaw repeatedly with grazing rights and left hooks. He frequently beat Gavilan to a left jab to the face. However, the earlier lead accumulated by Gavilan could not be overcome by this closing rush of the challenger.

Eddie Borden, with more time to reflect, wrote later in *Boxing and Wrestling* magazine and saw the fight differently:

> During moments of the battle, Billy used his famous left jab with the precision of a piston rod, and his right-hand punch to the body was a weapon which Gavilan found difficulty in avoiding. He used it steadily throughout the fight, and the Cuban apparently had no method to thwart that particular punch. Billy, too, looked good in landing. He pounded Gavilan's body continually, and it may have been responsible for slowing up Gavilan in the closing rounds.

More than fifty years later, Gil Clancy recalled: 'That night, well, there was a bar we used to go to, Gray's Bar – the owner's name was Eddie Gray – right near the old Garden. As soon as the bell rang at the end of the fifteenth round, I ran down to the place and said, "Eddie, we got a new world champion!" He says, "No you don't." I actually couldn't believe it, that's how bad I thought it was. It was a terrible decision.'

From that night onwards, Billy had the hearts of Manhattan. Forever. He was their champion. But he didn't have the title. He didn't have the bucks. He didn't have what he got into boxing for in the first place. Sentiment was fine, but it didn't pay the rent.

Whatever the brightness of Broadway's lights, whatever the magic in the air – as the old Drifters song tells us – there are shadows too, and the stench of failure and disappointment. That was such a night for Billy Graham. That was a night he didn't love New York so much, or boxing. That was a night he must have wondered why it had to be like this. He could take a fair decision; he could admit it if he lost. That wasn't a problem. But he found it tough to take that something beyond the power of his fists might have robbed him of a win he reckoned was legitimately his property.

The fans busted up bits of the Garden the night the Kid got the decision over Billy. It was not a major riot, but there were a lot of fists flying. People were angry, not just with the decision, but because they thought they'd been conned even before the first bell rang. Again. If the verdict was already in, what were they doing there shouting their heads off?

What the wise guys never understood, or simply ignored, was they were killing the game for short-term return. Fight after fight, year after year, of inexplicable results had a drip-drip effect on the business. Cynicism grew. Romance died. Boxing needed a saviour and one did not reside anywhere in the offices of Madison Square Garden.

But the shock that spoiled Gil Clancy's premature celebrations in Gray's Bar that night was as nothing to that which Johnny Sharpe experienced when watching Gavilan box Peter Waterman at Harringay Arena in London five years later.

Waterman, a neat and accomplished boxer from Clapham, had a good dig and would eventually win the British and European welterweight titles. His younger brother, Dennis, would go on to TV fame in *Minder*. But that February of '56, Peter was the Waterman the crowd had come to cheer – until the decision was announced. The referee and sole judge stunned everyone, including Waterman, by giving the decision to the Londoner.

Sharpe told readers of the *Ring*: 'I gave [Waterman] two rounds against Gavilan's seven, with one round even. Consequently, when referee Green raised Waterman's hand without any hesitation at all at the close, I wondered if I was seeing things.'

Two months later, in a packed Earls Court Arena, the Cuban won the rematch. And everyone went home happy. Just like Villemain had.

Joe Miceli was another of those pros who won and lost but was still regarded as one hell of a fighter. Joe had had quadruple bypass surgery, but, when I visited him in his comfortable bungalow in Brentwood, Long Island, he looked in decent shape for a man in his seventies. He was born Joseph Anthony Miceli in 1929 on Mulberry Street – the same street where Lucky Luciano was born and lived and schemed.

Boxing would not survive without fighters like Miceli. He was always there, as they say, always punching. He started as a prodigy, finished as a tradesman. And he had very few regrets. Along the way, Joe had some decent pay days, others not quite so well remunerated, and he got as close to the beating heart of boxing at Madison Square Garden as any fighter of his era.

He boxed for a living between 1948 and 1961, right through the decade in which the game haemorrhaged so much credibility it nearly died of shame. He won sixty fights, lost forty-two times and drew eight of his 110 bouts. That's about ten fights a year, with little time in between to spend the purse or have a good time. He was out of shape sometimes, and paid the price, but he didn't complain about it. 'What's the point?' he said. 'You take your licks.'

On the face of it, the stats don't indicate a remarkable career. But he worked in an era when fighters fought everywhere and anywhere, any time they were asked, for just about any money that was going. The conditions weren't always in Joe's favour, but he never turned down a fight.

Miceli loved the Garden. His first fight there, a Friday night in March of 1949, was his twenty-first bout as a pro. Joe had paid his dues all over town, winning mostly, in places like the Eastern

Parkway, the Ridgeway Grove, the Broadway, the Fort Hamilton, all venues famous in Brooklyn, and all gone now. So too are Croke Park and the Jamaica Arena in the Bronx, and St Nick's in New York City.

Joe came to the Garden well schooled and with a reputation for knocking people out with his left hook. He admits he might have been a bit 'left-hook-happy' – unlike his opponents, who often ended up 'left-hook-unhappy'.

This night, he gets an eight-rounder against washed-up Roman Alvarez, the semi-final to the main event, a ten-rounder between his friend Billy Graham and Brooklyn's Paddy 'Billygoat' DeMarco. Graham wins on points. So does Joe, outboxing Alvarez, who'd been around forever. Roman, also a Brooklyn fighter, was two days short of his twenty-seventh birthday but shop-worn and in the middle of a long, long losing streak. He'd been on the same circuit as Joe for nine years, losing to some great fighters along the way, including Kid Gavilan. This was his twelfth fight in the Garden, and his last.

Joe would go on to make the Garden his boxing home much as Roman had done. They were on boxing's treadmill. The difference was, many of Miceli's fights would be on television. He was to become a boxing star, one without a title but with a significant and enduring New York profile.

Miceli appeared in TV fights thirty-seven times. Only two boxers were on television more often in the fifties: Gavilan, forty-seven times, and Ralph Tiger Jones, forty-five. These were the faces Americans became familiar with when they were watching the fights on their new TV sets in the 1950s. Joe fought eight times at the Garden, three times in the main event. He won four, lost four. But the converted southpaw who started as a lightweight and finished as a middle will always be remembered by old fans – and those who had TV sets – for the power he carried in that left fist of his. Joe's trademark blow was a twisting half-hook, half-uppercut that arrived from nowhere and, twenty-eight times, put his opponent out for the count.

We open up his scrapbook, a wonderful library of highs and lows. The names tumble out as if he's reciting a who's who of boxing: 'The Keed', as Gavilan was known, world lightweight champion Ike Williams – an all-time great and no friend of the Mob, Wallace Bud Smith, who also won the lightweight crown, Johnny Saxton and Johnny Bratton, the welterweight kings, Don Jordan, who also won the title, as did opponents Curtis Cokes and Virgil Akins, and Gene Fullmer, the middleweight champion.

Joe was ranked among the top ten welterweights in the world for six years but couldn't crack it for a title shot. He takes pride too in having shared a ring with two fighters considered the best Cuba produced in that time: Gavilan at the start of his career, and Luis Rodriguez at the end.

We turn the page and there, browned but legible, is the newspaper account of his fight with Gavilan on 12 December 1950, Joe's first main event in the Garden. The decision was split: one judge saw it six rounds for Gavilan, three for Joe, with one even, another gave it 7–3 to the Cuban and the third saw it 6–4 for Miceli. Close enough to greatness to be well satisfied.

Miceli fought only twice more in the Garden: stopped in the eighth of ten by Johnny Bratton in 1952 when he was doing pretty well until decked for a second time in the seventh; and retiring after seven rounds against Charley 'Tombstone' Smith in 1958.

Joe's career was in decline when he dragged himself up the steps for the last time at the Garden against Tombstone, but he gave it a decent shot. There is no rancour as we leaf through his memories.

'Johnny Saxton, the third fight. Yeah, September '57. He quit after that, I think. And he died a bit later . . . Ah, Gene Fullmer. Three brothers – Gene, Don and Jay. All good boxers. Gene was a dirty fighter, though. Still alive. Not in great shape. He knocked me out. All rabbit punches. I beat his brother. My wife told him, "You're a dirty fighter. You shouldn't fight like that." He said, "I had to get even for my brother."'

Miceli doesn't differentiate between the wins and losses as we move through the book. They are all part of his life and he accepts them now as he accepted them then, even when he was losing on a regular basis.

'Hey! There's Yama Bahama, another tough guy I lost to. The nicknames they give 'em, I ask ya. He was from the Bahamas . . . But he was a good fighter.'

That was in 1958, Joe's tenth year as a pro. The left hook was doing it for him less frequently, but it was a living.

'I'm doing bad now,' he says, as the next report details another defeat. 'But I still fight, I don't care, win or lose. That's because I love it. I never thought nothing of it. I just made a few bucks. This is South America, now this guy he went down here, look at him. I hit him with a left hook. The first punch I went "Bang!" He goes down. He gets up. He butts me. They stop the fight. I lose. Six-rounder . . . Whatever. Max Baer was the referee. What a character.'

It was a ten-round main event, actually. But you get the impression Joe never expected to reach the final bell anyway. The opponent was Jorge Jose Fernandez and he'd won sixty-six of his seventy fights and was known as El Toro de Pompeya. Until the end of his career, which stretched on until 1973, El Toro hardly ever left Argentina, and rarely fought outside Buenos Aires. El Toro then went out in the big wide world of boxing and did pretty well. Joe was a stepping stone. He didn't stand a chance.

We turn the final pages.

'This is my last fight coming up.'

'What did you do that year?'

'I did nothing. I don't remember. That's my last fight.'

'This is against Curtis Cokes, right?'

'I was working over at the jail, I was a jail guard there.

Riverhead, out east, Long Island, way out. This is the last fight. I was never trained for that fight.'

'How good was Cokes?'

'He was like . . . in my eyes he was OK. But it was me who was washed up, in '61.'

We'd trawled his life, but Joe was happy to carry on reminiscing. The names – and that's all they are now – kept coming as we drank more coffee, ate more cake.

'Billy Wyatt, Kid Gavilan, Johnny Saxton . . . I knew all of 'em. Who was the best? Ike Williams.'

'Why is that, Joe?'

'Cos I beat him! He was a great lightweight champion. Nobody wanted to fight Ike.'

Miceli distils a dozen years of fighting into an afternoon and I'm wondering if that's fair on him. He doesn't mind, though. He likes to talk about the fight game, about the old days.

'See him? Sonny Boy West. Great fighter. I gave him such a beating, oh I don't believe it. He was such a great fighter and he went into the seventh, eighth round and he's coming at me with his hands down, coming right towards me. So I don't want to hit him. The referee said, "Hey, people paid money to see you fight!" "But the guy's half dead!" I said. I finished the fight quickly. I won big. The next fight, or soon after, in the ring he died. I felt I did it. I think Percy Bassett killed him. Could have been him . . . But he was shot to hell when I fought him.'

Sonny Boy West was a pretty good fighter. When he fought Joe, in September 1950 in Milwaukee, he was 43–6–1 and well placed in the division. He was a lightweight from Baltimore and also known as Al West. Up until 1950, he'd had a lot of easy wins around Baltimore and other out-of-town places but then came to New York to fight the dreaded Ike Williams at the Garden in February. Ike knocked him out in the eighth round.

Joe's memory has let him down a little, but the essence of his recollection is correct. Miceli carried Sonny to the final bell to record a points win. He just couldn't bring himself to finish him with his left hook. West then won a few low-key fights around Baltimore and Rhode Island before taking on Percy Bassett back in New York, at the St Nicholas Arena.

Bassett was good, too. Better than Sonny. When they met, five days before Christmas, he had forty-six wins and a couple of losses. Percy was hot and would go on to briefly hold 'interim' versions of the great Sandy Saddler's world featherweight title while Saddler was in the army. When Bassett fought West, he was up against a shell.

As one report had it: 'West complained of double vision between the sixth and seventh rounds. After he was hurt to the body by Bassett, he was floored by a right hand. As he fell, he landed hard on his head. West died of injuries suffered in this bout on December 21st. The official cause of death was given as a "inter-cerebral haemorrhage resulting from a cerebral concussion." '

But Joe still holds himself responsible. He saw several months and three fights earlier that Sonny Boy West had no business still being in the fight game.

'But it happens that way sometimes, doesn't it, Joe? Sometimes there's nothing you can do.'

'It is bad . . . I tell the referee stop the fight, and we continued to fight . . . It is bad.'

It was the only time in his career that Miceli backed off.

'Those were fighters them days. You had to fight your heart out always, and I did, I fought my heart out. But when I seen a guy getting killed like that I don't want to hit him no more. I felt sorry for this guy. His hands were down. Everybody saw him, he walked towards me with his two arms down. In other words it's, "I'm not quitting, hit me." I don't want to do that.'

'And you could see it in his eyes that he was gone.'

'Oh yeah, he was gone. A hundred per cent dead.'

The Korean War wasn't going so well and Miceli was drafted in 1951. But the guy who gave everything in the ring was not the most enthusiastic soldier Uncle Sam ever called upon.

'I was supposed to go to Korea, I went AWOL. They grabbed me. No, they didn't grab me; I turned myself in. They put me in the stockade for at least a month, then two MPs come and pick me up to bring me back to camp. When they bring me back to camp, the camp's deserted. "Where's everybody?" I say. "They all went overseas," they say. They had no place to put me, so they shipped me to Camp Kilman, New Jersey, and put me in there.

'I was fighting while I was in there. This kid Andy Candy Anderson. He was the champ of Indiana. So they made a fight with me and him. We fought a ten-round draw. I didn't like that. So the guy who promoted this fight grabbed me to fight him again in Grand Rapids, Michigan. I said, OK, good. So I tell my trainer. He says, "I can't be there at the start but I'll be there in time before you fight. I'll bring your robe up and your mouthpiece, everything." And I wait for him. Fight time: "Joe Miceli, get ready." I can't get ready, he's got all my stuff. So I borrow a pair of shoes, I've got nobody to work with. I get up a dollar mouthpiece and I tell one of the fellas and they were, "Yeah, sure, we'll do it." So I go out there, round one: bang! Round two, knocked him out. Now I go down to the dressing room and get ready to dress up. Here he comes, my trainer, with the suitcase, says, "Joe, get ready for the fight." I said, "The fight is over. I knocked him out." Ha!'

Miceli was a no-fuss fighter in every way. He is warming to his anecdotes now. 'Anybody. I'm number one. Fought more black fighters than anybody. I'm number one. That's the way it is. Friday nights I used to fight at the Garden. Wednesday nights out of state on TV, and I had four on Saturday nights. Clubs, whatever. Yeah, all round the States. And I loved it. I loved every part of it. I fought anybody. To me, the fighter he's got two arms, two legs, like I've got two arms two legs. They all were good fighters them days. Every one of them were good fighters. Even if they had no name, they were good fighters. They stood there, ten rounds. The TV paid money, but not that much. I forgot how much it was. Three thousand. Best punch left hook, uppercut. One-Arm Miceli they call me. Cos I never hit with the right hand. No power, nothing.'

In all his fights, Joe Miceli knocked out one opponent with his right hand – and he still can't quite believe it happened.

'Me to drop a guy with a right hand? I knocked him down and he hurt his leg. Ramon Fuentes.'

'You didn't believe him?'

'Of course not. It's only advertising.

'Bobby Nelson was my first manager. He sold my contract to Harry Stickevers in 1953. Don't know why I never got a title shot. My manager gave me a black name. Give me a bad name in boxing, my manager, cos he sold my contract. If I'da stayed with Bobby, I mighta got a world title shot. I dunno. He called me yellow, my manager, Bobby. Why did he call me yellow? For me to get mad? I got mad and I got outta my contract. Bobby, he thought I was lazy. I *was* lazy and I didn't do it good then. He didn't like the way I used to fight. I used to get tired at the end all my fights. I'd go ten rounds and I'd lose the fight because I got tired. My manager told me, he said, "Nobody will beat you, nobody, if you train like a fighter." I thought he was conning me and all that . . . but that's the way it was.'

Joe got out of the army and got married. The scrapbook shows a lovely dark-haired bride and a smiling groom, who looks slightly out of place in his smart suit.

'There's when I got married, September 26 1954. Catherine Tuzzo. That's her maiden name.She lived on Mulberry Street. We had a big wedding. Football wedding, as it was called. Everybody came with sandwiches around them.'

Joe set about earning a living to put food on the table for Catherine with the only tools he ever had. He beat Bud Smith –

The beginning of the end: Rocky Marciano, who knew plenty of Mafia guys but was never implicated, lands the knockout blow in October 1951 that finished Joe Louis's reign. Joe retired and lent his faded name to the IBC's takeover of Madison Square Garden.

An unlikely opponent: Senator Estes Kefauver from Tennessee pursued the crime tsars throughout the fifties with quiet determination, nailing some of them finally in 1960.

Hands of the devil: Crime boss Frank Costello begged not to have his face on screen when Kefauver began his televised investigations. Kefauver wrote in his account of the hearings that, 'The nails of his nervously writhing fingers on the hairy-wristed hands were freshly manicured

The righteous writer: Estes Kefauver, a presidential hopeful of the early fifties, capitalised on his celebrity when he wrote his surprisingly lively 1951 account of the first Senate investigation into organised crime. The boxing community read it with interest.

Gone bad: Benjamin 'Bugsy' Siegel, the handsome mobster who built Las Vegas, was quietly reading the newspaper at home in Beverly Hills in June 1947, when blasted into oblivion by the gangsters who were once his partners.

Retiring figure: Frank Costello claimed to have got out of the rackets, and boxing, when posing for this relaxed shot at the Beverly Club in New Orleans in March 1949. There was much to come from him.

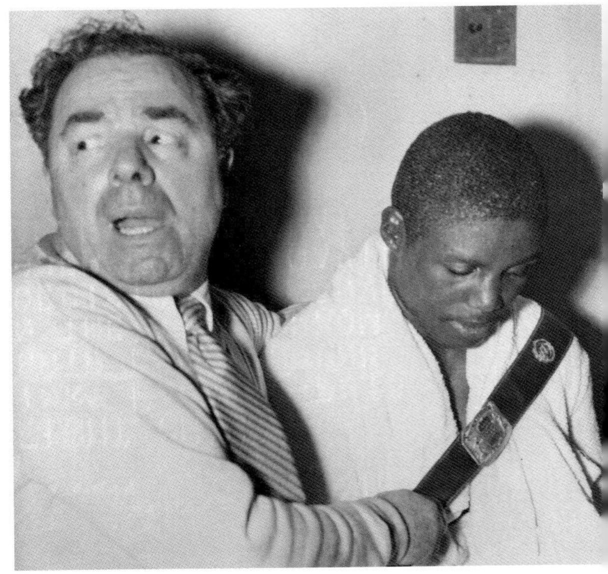

Blink and you'll miss him: Blinky Palermo allowed himself brief public exposure to put the world welterweight title belt around his fighter, Johnny Saxton, after he'd beaten Kid Gavilan in October 1954. Gavilan reckoned the fight was fixed.

Gentleman Jim: James D. Norris looked anything but a gangster and, seen here at the fights in 1951, added a veneer of style to the doings of the International Boxing Club.

Another face: Truman Gibson, like Norris, strove for respectability but was eventually caught up in the IBC net. He looks calmness itself trying to convince the US Supreme Court in January 1959 that the IBC was not a monopoly. It didn't wash.

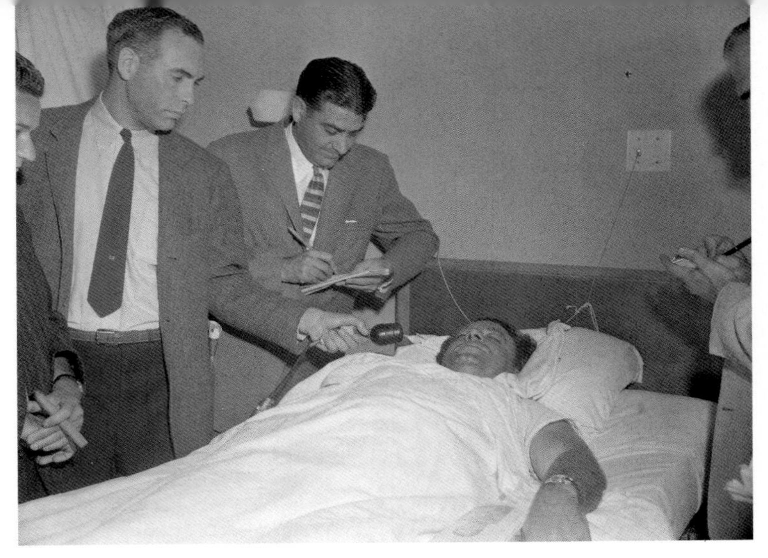

Knockout: Reporters gather around the hospital bed of Los Angeles promoter Jack Leonard in 1959 after Blinky Palermo had him beaten up for not 'cooperating' in two world welterweight title fights between Don Jordan and Virgil Akins.

The End: Frankie Carbo, Mr Gray, leaves court in 1959 after being prosecuted at last for masterminding the almost total manipulation of boxing in the fifties.

Tell me about it: Don King has always denied Mob connections. Nevertheless, he eventually slipped seamlessly into the vacuum they left to become the most influential figure in boxing for the past forty years.

King of Nothing: few world heavyweight champions have been so ruthlessly manoeuvred as Charles 'Sonny' Liston, pictured with his paper manager, Pep Barone, a front man for Blinky Palermo.

Moving on: Liston, dethroned in suspicious circumstances by Muhammad Ali, almost certainly threw the rematch, flattened by a punch few saw in round one in Lewiston, Maine, in May 1965.

twice – drew with Joey Giardello, outpointed Virgil Akins, gave Gil Turner a scare, upset a rising, unbeaten Stan Harrington, hung in against Ralph Dupas, so confused Saxton with the power of his punch he went to the wrong corner at the end of the third round and was hauled out, put Art Aragon down twice on the way to a points win nobody expected . . . but the upsets stopped coming after a while. The beatings got worse. Joe got out about the right time.

Stickevers, tall, suave, smartly dressed and softly spoken, was an influential man in the Managers' Guild in New York, but Miceli never knew if that was a good or a bad thing. He didn't ask questions.

'Did you meet those guys, Joe, guys like Carbo and Palermo?'

'Oh sure. I knew 'em. Carbo, Palermo. They were as big as Don King is today, that's all.'

'Nice guys?'

'I think Frankie was a swell guy. He was friendly, just that he was with the Mafia. That's all. I didn't bother them, they didn't bother me, that's all.'

'Did your manager have to be in the guild for you to get all those TV fights, Joe?'

'See, you had to be in the guild to fight. I wouldn't have got the TV fights, no. You're right. So he paid the guild.'

'And everyone was OK with that?'

Joe doesn't answer. He turns to the scrapbook and opens it up.

'See him? Fought him . . . and he bought it. It was three tens in the Garden. January 1950.'

Miceli had two fights that month, both in the Garden. The first was against Buddy Garcia, from Galveston, Texas. Joe knocked him out in the first. Buddy had one more fight and quit boxing. Two weeks after Garcia, Miceli fought the Cuban Raul Perez, who'd not long got off the boat from Havana. Joe stopped him in the seventh round of a ten-rounder, just like the cuttings in front of us said. And Joe thinks Perez 'bought it'. The records show he pulled out at the end of the seventh with a hand injury. Perez hit the road for a while after that loss, but he would turn up again at the Garden, two years later, to get knocked out in the first round by Johnny Saxton. And Johnny was very well connected.

'Do you think there were fixed fights in those days, Joe?'

'Fixed fights? It's the fighter himself that makes a fix. They're not fixed, otherwise. You understand? If I'm fighting you and I'm the big three or four to one, I'm fixing myself to lose. Now nobody knows. You understand what I'm talking about? Heh, heh! Maybe ya don't. Heh, heh! Hey, Catherine, bring our guest some more o' that damn cake, will ya?'

The sun is going down. Joe looks tired. I get ready to go. The only thing I hope Joe ever threw was that left hook. It must have been a beautiful sight.

11

SONNY AND THE MOB

Joe Miceli was just making a living, doing the best he could. But Estes Kefauver had little time for fighters' hard-luck stories. He viewed the professional wing of the sport as an aberration, a vulgar outgrowth of man's baser instincts. But he was determined to clean it up, just the same.

The catalogue of suspect fights, rumours of arm-twisting hoodlums forcing managers to sell pieces of their fighters to silent partners, dwindling crowds and a general sense of disenchantment spreading through the fight game, even in its hardcore constituency, convinced the senator the fight game was rotten right through, like a good apple that had been left in the sun too long.

But his first investigation into organised crime had proved largely inconclusive, at least legally. Carbo, Costello and all the other suspects, advised by the smartest of lawyers, simply played dumb at his inquiries. When he took his investigation – called, in full, the Senate Special Committee to Investigate Crime in Interstate Commerce – on the road, bad men did not exactly tremble in their spats. They hit fourteen cities and took evidence, if it can be called that, from about six hundred witnesses, many of whom blanked the senator with practised insolence.

The roll-call of criminals Kefauver quizzed was impressive enough. As well as Costello and Carbo, he asked tough questions of the gambling boss Willie Moretti, who was Costello's cousin and a member of the Genovese family, and Joe Adonis, who was born Giuseppe Antonio Doto and who deserves credit, at least, for choosing a striking new name. Joe was a petty thief, pimp, rapist and an associate of Lucky Luciano. After his walk-on part in the hearings, he was eventually deported back to Naples.

Thwarted by the Mob, Kefauver decided on an easier assignment: running for president. He put himself forward as a candidate in 1952 and, to the amusement of many, he hit the hustings wearing a coonskin hat, in homage to his Tennessee roots. The Keef won twelve of fifteen primaries, including a shock result over Harry S. Truman, but lost out to Adlai Stevenson on the convention floor. Stevenson pipped him for the nomination again in 1956, but he ran as his vice-presidential candidate – to no avail, as it happens.

So, his ambitions to lead his country lay in ruins. A lesser individual might have walked away from politics altogether and resumed a lucrative calling at the Bar. But zeal still burned in Kefauver's heart. He was still on the Mob's case. One way or another, he reckoned, he would get the bastards.

For the International Boxing Club, meanwhile, business was not so much as usual as weird. Their monopoly of the business was so great it was embarrassing, even for them. Between 1949, when they captured Joe Louis and his heavyweight title, and 1955, when the investigators came calling again, the IBC staged forty-seven of the fifty-one world title fights held in the United States, all of them in their own stadiums.

Norris and his Chicago business guru, Arthur Wirtz, were the nominated frontmen of the IBC and were now putting on fights at the Garden, the Polo Grounds, Yankee Stadium, St Nick's (all in New York), Chicago Stadium and the Detroit Olympia. Norris and Wirtz had a piece of all the venues, so they were promoters and site operators all in one. It was this stranglehold on boxing that kept

Kefauver's investigative instincts honed. If he could not prove out-and-out racketeering and coercion, he would get them on a legal technicality.

Encouraged by Kefauver, Judge Sylvester Ryan of the US District Court finally ruled that the IBC constituted a sporting monopoly. By this time Norris had tried to manoeuvre his way out of the prosecution spotlight by resigning as president of IBC, New York, and withdrawing his interests to Chicago, his home base. In his place, Gibson stepped in as president. So, the lawyer who'd started his working life campaigning for the rights of his fellow blacks in the armed forces was catapulted to the head of an organisation that did as much as any other in sport to keep them in subjugation, to work for criminally poor purses and to answer to the whims of the Mob, or risk exclusion from the only trade they knew. It was some journey.

Judge Ryan was unimpressed, either way. He saw the ruse for what it was. He knew that Norris was still, effectively, the boss of the New York operation, given the amount of trans-state television shows emanating from the place. This is what did for the IBC, eventually. Because those fights were generating income in other states, the company staging them, the IBC, was considered to be a monopoly.

Then Gibson and Wirtz were, laughably and to the disgust of those people in boxing who wanted to see the back of the Mob, given five years to wind up their operation. This was of a piece with all the slap-on-wrist fines and suspensions the New York State Athletic Commission handed down to the likes of Graziano and other mobbed-up pugs, who could afford to laugh off the penalties as mere inconveniences.

The judge's ruling was more punitive in one telling regard, however; it had to be carried out eventually. What Truman Gibson was in charge of from about 1955 onwards was virtually a fire sale of the IBC's assets. Their days were numbered. Kefauver didn't make it to the White House. But he had finally landed a blow on the crime tsars who ran boxing. The Chicago operation was also declared a monopoly. Ever so slowly the bad days were drawing to a close. And there would be more devilment before Carbo, Palermo, Norris and Gibson were finally driven out of the fight game.

The four of them had friends, still. Some were friendlier than others, of course, but there was the unspoken code of the business to be respected: what worked stayed in place. This was business. It was business as it had always been conducted. There were too many interested parties deeply involved in the old ways for them just to pack their tents and leave the campsite.

Then, on to the scene, rumbled a most menacing figure, a man like no one previously in boxing, a fighter so fearsome he scared opponents before they got anywhere near the ring. And he was mobbed up to his eyeballs. His name was Sonny Liston.

The heavyweight champion by this time was a fighter Red Smith of the *New York Times* called 'the man of peace who loves to fight'. That wasn't entirely true. Floyd Patterson, a God-fearing Catholic who survived reform school, was an introvert who found peace in the ring, not as a pure 'fighter' but as a boxer. He was a slender, athletic technician with astoundingly quick hands and fists of dynamite. His every physical move was at odds with his nature. His trainer, D'Amato, the man whose paranoia about the Mob was monumental, said of Floyd, his first and favourite champion, 'He was a kind of a stranger in this world.'

'You can hit me and I won't think much of it,' Patterson said once, 'but you can say something and hurt me very much.' It echoed the sentiments of Joe Louis, whose view was, 'You don't have to hate somebody to fight them. No need to hate nobody.'

Patterson was born in a cabin in North Carolina on 4 January 1935, the year Joe Louis signed with Mike Jacobs. In fact, that Friday night, Joe was beating Patsy Perroni over ten rounds in the Olympia Stadium in Detroit.

Nobody is sure when Charles Sonny Liston was born. It might have been 1928, as his mother maintained, or it might have been 1932, as Sonny said. What is known is he was born in even humbler circumstances than those of Floyd Patterson.

His story is a remarkable one on many levels, and illustrative not only of the condition of his people in those tough times but of how fighters remained virtually powerless, whatever their physical aura – which, in Liston's case, was considerable.

Like Floyd, he was quiet, contemplative even. But he was a brooding and injured man, who did not seek affection because he suspected it would be futile, and so, like many young black men in the same situation, then and now, he used aggression as his shield.

In the course of researching an article for James Brown's now-defunct *Jack* magazine a few years ago, I stumbled across some revealing facts, and a little fiction, about Liston. These are some of them . . .

Riley King, who started out as Beale Street Blues Boy Riley King in the bars of Memphis in the forties, stirred deep, soulful feelings among listeners to the local black radio station that gave him his first major exposure. He shortened his name to Blues Boy, then B.B., moved to New York and played his guitar, which he called Lucille, so sweetly everybody came to call him the King of the Blues.

He sang this, often:

> Bad luck is fallin'
> Fallin' down like rain.
> No matter what I do
> Seems like my life
> Won't ever change.

And one man from the Deep South who listened carefully to every word and every note B.B. King sang and played was Sonny Liston. He had good reason to. They had history.

Sonny was the second youngest of twenty-five children that a rat-thin cotton-picker called Tobe Liston had by two women, the ninth of ten he is said to have had with a fine, big article called Helen. She had another child, her first-born, E.B., by a Mr Ward. Mr Ward, however, did not stay around long, leaving E. B. to be brought up by Helen with Sonny and the rest of the battalion. When E.B. married, it was to the mother of B.B. King. So Sonny Liston, who made fighters wet their pants in fear, was half-brother to the stepfather of the finest blues guitarist most of us have ever heard. This effectively weakened a myth that Liston was actually King's uncle, but then these stories circulate like bad air in boxing.

I discovered that Mike Tyson knew this piece of black boxing history, and he knew all about Sonny. He was big on history, whatever the perception of him as an illiterate thug. It is a stigma he shared with Liston. Mike knew that Liston was born in the early years of the Depression, or maybe just before the Wall Street Crash of '29 – not that either of those cataclysmic events would have materially altered his circumstances. Sonny was born into shoeless poverty in a hut with no roof, hidden among a hundred cotton fields in Arkansas. Helen couldn't be sure of the year, only that it was January. 'It's cold in January,' she said.

Sonny sometimes said he was born in Pine Bluff, which is thirty-four miles from Memphis, across the state line in Tennessee – Kefauver's state. His mother said no, it was a place called Sand Slough, Arkansas, and it's not on any map.

Wherever and whenever he was born, Sonny found trouble without looking. When he stole the bus fare and went to St Louis, Missouri, to find his mother (she'd had enough of Tobe and his wild ways) there was no cotton to pick, but plenty of pockets.

With a couple of new friends, Willie and James, Sonny rolled drunks and was entered on his first rap sheet as 'No. 1 Negro'. It was the nadir of anonymity. Arrests and charges for petty crimes did not deter him, though. This was his identity. White society had made it so, and he was content with it, given the lack of options. Sonny kept mugging, went to the penitentiary. Inside, a Catholic priest claimed he taught Sonny to read, write and box. When he came out after five years, Sonny couldn't read a bus ticket but he had a jab that could kill. He was fixed up with union racketeer John Vitale, whose confederates were Carbo and Palermo. He didn't know it, but from this point on, Liston's life story was already written down. In 1953, Sonny went to Boston for an amateur boxing tournament and raped a hotel maid. A call from St Louis fixed it. Now he was in hock to the Mob forever.

Sonny's almost superhuman straight left was his in-built vibrato and he started making serious noise with it. But he'd have to wait to crash the upper reaches of his trade. He was too elemental a force in every way – for opponents, for the smart white writers who loathed him, for the president, even. Glory was down the road.

> Bad luck is fallin'
> Fallin' down like rain.
> No matter what I do
> Seems like my life
> Won't ever change.

Liston's professional debut was a perfunctory affair, and it would be in St Louis, the home town of the criminals who owned him. Sonny was pitched in against a small but formidable heavyweight from Louisville called Don Smith, who could hit but couldn't ship a punch himself. Liston mugged him. It was not conventional boxing. So wild was the encounter the referee Jimmy Parker found himself powerless to break them up when Liston engulfed his victim in a storm of thick-armed swipes, ramming him up against the ropes and turnbuckle as if Don owed him money. When Parker did separate them, Smith was practically unconscious on his feet and had a deep cut across one eye where Liston had lacerated the skin with the laces of his gloves. The Mob had their monster.

His education, if that is what it can be called, continued in a similar vein around the rings of Missouri against frightened opponents thrown in like meat to a rabid dog. Liston's extraordinary natural strength was too much for all of them – until he came up against a smart operator called Marty Marshall at the Motor City Arena in Detroit in September 1954. Liston had posted seven wins, two of them awkward split decisions over cagey opponents Benny Thomas and Johnny Summerlin.

Marshall was a rough customer from Detroit, ten years older than Liston and having his twenty-sixth fight. A few months earlier in Columbus, Ohio, he'd fought and beaten the same fighter, Herbie Moore, twice in an evening. He knocked Moore out in three rounds, and an hour later the promoter Al Haft paid them to do it again. This time Marshall did him on points over ten rounds.

That was the level of fighter Sonny was up against – and he felt Marshall's considerable power in the fourth round when a left hook broke his jaw. But he hung on to go down narrowly on points over eight rounds. It was a setback, but not a major one. Liston resumed his marauding of the division. His handlers were pleased enough and would 'move' him when they got him into the right position. The one obstacle they had to overcome was the stubborn D'Amato.

José Torres, who won the world light-heavyweight title for Cus in 1965 by knocking out Willie Pastrano in the Garden, is a complex and unusual fighter, a Puerto Rican who writes beautifully and with insight in English. He was with D'Amato in the fifties when Liston was scaring the entire fight game with his aura and his power.

'I remember Cus saying to me once – and I agree with this, utterly – that the art of a promoter, manager, the guy guiding a fighter's career, is to protect his man at all times. Matchmaking, he said, was about making sure your guy won. Those people who say Cus protected Floyd when he was champion are right; that's what he was supposed to do. This is a business. It's not just for the entertainment of people in the crowd. It's a boxer's living. The title is the way he makes his money, and nobody knew that better than Cus. He'd worked hard to get us into that position and, when he saw Liston, he did not want a part of him. Part of it was because he didn't want to do business with the people around him. Part of it was he didn't want Floyd to lose his title. How can you blame him on either count?'

This does make sense – from D'Amato's view of the world. It in no way takes into account that Liston might deserve a chance to take Patterson's title. The title, after all, was not just Floyd's. It belonged to boxing. There was a lineage stretching back to John L. Sullivan, admittedly not always entirely creditable, or credible. In his morally firm way, D'Amato was doing what the Mob were doing in theirs: he was hijacking the world heavyweight title, for as long as he could get away with, for as long as he could keep Patterson in place to make more money with the belt. He knew his weaknesses, and, more importantly, he knew the savage strengths of Liston.

So Sonny waited. And waited. And won. And won. The years passed slowly. Liston fought where he was told, from Duquesne Gardens in Blinky's town, Pittsburgh, on through Victory Field, Indianapolis, the Auditorium in Miami Beach, where hoodlums got their sun, back to Missouri, on to Chicago and down to Texas, knocking them out as he went, good fighters, seasoned and dangerous pros such as Cleveland Williams, Willie Besmanoff, Mike DeJohn, Nino Valdes. Year after year, Liston toiled. And year after year, D'Amato turned him away. The ratings panels and the rascals at the *Ring* and in the papers who conspired with them to vilify Liston snuffed out his legitimacy even as his awesome reputation grew.

Meanwhile, D'Amato steered Patterson towards stiffs, old guys and barely believable challengers. He started his reign by beating up old Archie Moore in 1956. The Ol' Mongoose who turned up in Chicago that November brought the bruises and memories of a phenomenal 159 wins, twenty losses and nine draws into the ring with him. He was, he said, thirty-nine – sixteen or so years older than Liston. Floyd stopped him in five. Then he took out Tommy Jackson in ten at the Polo Grounds in New York before indulging himself in perhaps the most bizarre defence of the world heavyweight title ever contrived.

Peter Rademacher had won the Olympic gold medal in 1956. This, as hard as it is to believe, was his professional debut: challenging for the world title. Nevertheless, he managed to put the champ down in the second – before taking the trip himself six times, put out of his considerable misery in round six. The referee at Sicks' Stadium in Seattle was Tommy Loughran, a genuine ring legend; what he must have thought of the spectacle as Rademacher bounced up and down to the rhythm of Floyd's left hook is too bleak to contemplate.

Rademacher, a genial man from Grandview, Washington, boxed on for a further twenty-two contests, rarely disturbing the fight firmament as he eked out fifteen wins and suffered six knockouts (including one by Brian London at the Empire Pool, Wembley, in 1960) in seven losses. It was an undistinguished if interesting career.

And it underlined the absurdity of the Torres philosophy: as Liston fumed, and Carbo and Palermo grew increasingly irritated with D'Amato, the world heavyweight title was being demeaned.

Estes Kefauver was getting impatient. He had watched, powerless, as the IBC moved with smug assurance out of his clutches. But he was determined to have one more go at them.

The FBI, meanwhile, were on the case. They had tracked Liston for years, watching and listening to his every contact with Carbo and Palermo and Vitale. These were the guys they wanted to nail; Liston was just the blip they kept seeing on the radar.

The bureau's file No. 163-1275 is headed 'Subject: Charles "Sonny" Liston' and it is a cold, driven document. It presupposes guilt at every turn, often with justification; Sonny was no angel. He'd been hounded by the police from Denver to Philadelphia and all points in between. It made him more morose and irascible than he already was, feeding his sense of persecution yet having little effect on his performances in the ring. That is where he felt most comfortable, inflicting legal pain.

His rap sheet was long and ugly. The first entry reads: 'June 1, 1950 – plead guilty robbery first. Means deadly, dangerous weapon (2 charges), robbery first larceny from the person (2 charges) – 5 years, October 30, 1952, paroled.'

There followed updates as if on a school report, except detention for Sonny was more than doing lines:

January 28, 1953 – investigation – released January 29, 1953.
May 5, 1956 – assault to kill – January 28, 1957, plead guilty. Sentenced to 9 months in city workhouse.
June 21, 1956 – investigation.
October 4, 1956 – investigating suspicion larceny – released on charge of investigation suspicion stealing.
September 28, 1957 – strong arming – no entry.
August 12, 1959 – suspicion gambling – no entry.
June 12, 1961 – impersonating an officer – no entry.

Even when Sonny tried to move outside of their immediate jurisdiction, the FBI employed agencies in other countries to track his moves. One such entry in his file is logged thus, from their agent in Ottawa: 'Information urgently needed as Liston has apparently applied for permission to come to Canada for a possible heavyweight fight in near future. Sucab. 11–15–61.' He was subsequently picked up on a string of fiddling misdemeanours, none of them brought to prosecution.

And on it went, month after month, year after year. His every moved was scrutinised, some more diligently than others. An internal FBI memorandum of 30 April 1962 revealed the following:

> The Director asked, 'What about this?' in connection with the statement by columnist Dan Parker in the *New York Mirror* newspaper of 4/28/62. Parker, in his column, stated, 'It is known that only two days ago FBI men were in New York inquiring about certain Philadelphians who have been active in behalf of Liston.'
>
> Our New York office advises that no investigation has been conducted concerning Liston, and no contacts have been made by agents of that office with the State Boxing Commission concerning this matter. The New York Office advised that the State Boxing Commission has been making inquiries, but no known investigation has been conducted on a Federal level in New York City.

No, there had been no official inquiry. But, even when they weren't actively seeking dirt on Liston or 'certain Philadelphians' they were watching him. This was Big Brother writ large, as revealed in the closing paragraph of the memorandum:

> Our criminal intelligence microphone surveillances have picked up considerable information regarding Liston, which we have furnished to the Department. A summary is being prepared on Liston.

In New York, Frankie Carbo was getting his own grilling. The District Court inquiry, according to documents, had this to say of Mr Gray:

> Carbo, with a background of underworld association, emerges as the leader of the coconspirators. Gibson's first contact with him came shortly after the boxing clubs were organized some time in 1950. From time to time between 1954 and 1957, Gibson caused payments amounting to approximately $40,000 to be made to Carbo by the boxing clubs through the device of placing Carbo's wife, Viola Masters, in the club's payrolls for fictitious employment. Payment was explained by Gibson as being for the purpose of securing Carbo's goodwill and of preventing fighters or managers with whom Carbo had 'influence' from becoming antagonized or alienated. In investigations conducted by the United States Senate, Gibson had explained why Carbo's wife had been employed rather than Carbo himself: 'Because it looked a little better on our records, not ever considering the possibility of being called before a senative investigative committee, to have Viola Masters down instead of Frank Carbo.'
>
> Both before the Senate Committee and in trial, Gibson acknowledged that in their operations the boxing clubs had had dealings with the underworld. This practice was resorted to in order 'to maintain a free flow of fighters without interference, without strikes without sudden illnesses, without sudden postponements.' Gibson testified that the boxing clubs would use 'everyone' they could to prevent fixed fights and that in 'everyone' he included the underworld.
>
> On redirect examination, Gibson was asked whether it was 'the policy of the International Boxing Club or any

> other organization with which you were connected in the fight business to use force or violence or threats of force and violence,' to which Gibson replied, 'No indeed.' On re-cross examination he was asked: 'Of course, Mr Gibson, when you used various people to achieve your ends, you didn't know what those other people were doing on your behalf, did you?' To which Gibson answered: 'Not completely, no.'

And that about summed up the whole damn story, from start to finish: 'Not completely, no.' In one pithy, considered reply, the IBC's legal man, their respectable front, had described that area of uncertainty in all their dealings. To do so required that he not ask hard questions. If the operation were not to fail, he had to make a pact with his conscience – because, while he might not have known all the details of the skulduggery, he suspected what they were. He might even have been told what they were and, by some chemical trick of the brain, forgotten them. For corruption to thrive, there has to be conspiracy. And conspiracy, by definition, demands a common purpose. That shared goal was the accumulation of wealth and power. It had nothing to do with titles, really. None of them much cared who was champion. If they did not have the No. 1 man they would go out and buy a piece of him. They would then control both him and his manager, as well as the title. And it was the IBC who determined when, where and against whom the champion would defend that title. They might have been Romans organising the deaths of the Christians at the hands of the lions. Among the bravest of those lions was Sonny Liston.

While the FBI heat was on Liston, the International Boxing Club turned their attentions west, to Los Angeles. And this is where they came unstuck, through hubris, idiocy or plain arrogance hardly matters. The tough guys moving the puppets by now thought they were untouchable, whatever the ruling of Judge Ryan in 1955. They would carry on for as long as they could, lying, manipulating, bullying. They were milking boxing dry with their outrageous demands and sheer disregard for what remained of the sport's conventions. Carbo and Palermo were so accustomed to their booze and gambling lackeys doing as they demanded, they expected fighters, managers and promoters to respond with similar meekness. And they did.

The fight that proved their undoing, however, was not a world heavyweight title in the Garden. It was not an event of such magnitude that it required strong-arm tactics. The returns were healthy without being eye-bogglingly large. They were brought to grief on a petty rock.

12

THE SLOW DEATH OF THE IBC

Just before Christmas of '57, Carbo, emboldened now (in company with Costello, Palermo, Norris and, probably, Gibson) by surviving the attentions of Kefauver, Judge Ryan and the unsubtle wiretaps of the FBI, called a meeting of his cohorts in Miami. In attendance was a representative of a gathering the previous month of the Mafia's various families in the Appalachian Mountains. Many of the leading managers in the fight game assembled to listen to what Carbo and his guest from the inner sanctum of American organised crime had to say. What Carbo told them, on one particular issue, involved the immediate future of a so-so fighter from St Louis called Virgil 'Honey Bear' Akins. Mr Akins, Carbo informed them, was to be steered towards the world welterweight title recently given up by Carmen Basilio, a client of his friend Blinky, and, to facilitate his smooth progress, he would be 'placed' in various fights that would be judged to his advantage.

They drew up a list of opponents and presented it to Jim Norris, who rubber-stamped the edict and passed it on to the appropriate state boxing commissions for their unquestioned approval, which was duly forthcoming.

Akins, as per instructions, defeated one Isaac Logart in the Garden at the very time, *Sports Illustrated* later reported, 'DA Frank Hogan's men were passing out subpoenas in the crowd to start an action that resulted eventually in Carbo's indictment as an undercover manager'.

This did not bother Frankie. He was preternaturally calm about the attentions of the police, as ever. He would carry on in the manner he had grown accustomed to, that is to say as if he were the king of the world. The Akins plan was to be followed to the letter. On 6 June 1958, Virgil knocked out the accomplished Vince Martinez in St Louis, Sonny's town, as planned, and became champ. It was a result out of kilter with the available evidence about their form. Akins at the time had a mediocre record of forty-seven wins, eighteen defeats and a draw; Martinez was a stylist from Jersey with decent power, a big following at the Garden and a lot of good names on his log: Chuck Davey, Chico Vejar, Kid Gavilan (twice) and Gil Turner. When he went in against Akins, Martinez's record was sixty and five. He went down five times in the first, and twice each in the third and fourth rounds, when it was stopped.

As *SI* informed their readers: 'That night, the St Louis police intelligence squad picked up Blinky Palermo, a Carbo errand boy, and found him carrying an assortment of sleeping potions, including Seconal. At the time there was no special reason to believe that Blinky was using the drugs for any purpose other than, as he put it, to ease his aching back.'

However, Akins was not to reign as long as expected. There were complications. The IBC had not counted on Don Jordan, who was trained by the venerable Eddie Futch, and was a big name in California, spoiling their long-term plans when he was engaged to challenge Virgil six months later. Palermo had specifically warned Jordan's manager, Don Nesseth, that it was not in his interests to let his man perform to the utmost of his abilities. So, unable to persuade him to actually throw the contest, Palermo demanded that he give up 50 per cent of his interest in the fighter, and they would let the cards fall where destined. But Nesseth was not interested. Neither was the promoter, Jackie Leonard.

Leonard called Gibson, pleading for him to intervene with his silent partners. Gibson denied there was such a strong link that his intervention would have much influence but he said he would try. He knew it was pointless, of course. The psychopathic Palermo was not a man who would be dissuaded from any course of action he deemed was in his own best interest, and in this instance, his overriding concern was to keep control of the welterweight title. And that meant owning Jordan, like he had owned a string of fighters all over the country with whom he had only the flimsiest association.

Gibson's advice to Leonard, however, was unconventional. He said he should assure Palermo he would cooperate and then just see how the fight turned out. Leonard took this as a clear and dangerous hint that he could cross the Mob. He did not feel disposed to do so, but Gibson told him in a celebration of glibness, 'Jackie, that stuff went out with high-button shoes.'

Gibson went back to Palermo and told him the deal was in place. Leonard, however, proved a brave, if nervous, adversary for Palermo and stood firm: he was not going to give in, whatever the consequences. The thugs were briefly confused, as they were not used to such intransigence. But they knew what to do. They inundated Leonard with threatening phone calls. Palermo was joined in the wave of intimidation by Carbo and a couple of hired bruisers, Joe Sica and Lou Dragna.

Leonard went to the police, figuring he would flush the mobsters out, or deter them by the fear of prosecution. He knew not with whom he was dealing. The police told him to be careful, to take no risks. In short, they were advising him to play ball, or take his licks. That is what happened. As he was closing his garage door one night, the thugs struck, hard and heavy, beating him about the head. Then they fire-bombed his house. A stranger harangued Leonard in a public park, in broad daylight, screaming 'stool pigeon' at him. Bums on skid row were offered $250 to beat him up if he came by. The Mob wanted to make Leonard's life a nightmare and they just about managed it.

The fight went ahead, at the Olympic Auditorium in Los Angeles, a few weeks before the Christmas of '58, and Jordan won handily. There was not a sniff of interference – in the ring. Palermo had been insulted and he would take his slice of meat, one way or another.

As it happened, however, the FBI were moving in on the participants in the Mob summit meeting in the Appalachians, and they found a crossover with the heavies in boxing, namely Carbo and Palermo. The Jordan–Akins fiasco had arrived at precisely the wrong time for the IBC.

So city, state and federal authorities moved as one, a phenomenon. The California State Athletic Commission got the ball rolling by instigating an investigation into the machinations behind the scenes of the Jordan–Akins bout, which was under their jurisdiction. The Los Angeles Police Department finally got mobile, through their intelligence unit. And the boys from the Bureau came to town.

Things were looking grim for Carbo, Palermo, Norris and Gibson. They were all implicated. Such are the pitfalls of confederacy. One in, all down.

It would not be entirely straightforward, but the investigators for once showed gumption and went after their prey with glee. They smelt blood – and, this time, it didn't belong to a poor sap in gloves. They rounded up Gibson, Carbo, Palermo, Dragna and Sica and charged them variously with 'conspiracy to violate the federal Anti-racketeering Act, interstate communications, extortion and conspiracy'. Gibson, the lawyer, stood accused only of conspiracy. He looked like he might get out of it with a serious ticking-off, perhaps be debarred but not have to go to prison. The others had not even the thinnest veil of respectability behind which to hide. They were doomed because of their low standing in society. How deliciously paradoxical that it was only the black man among them who escaped prejudice.

The FBI took over from the local cops and carried the case to its logical, and moral, conclusion.

This one, seemingly trivial, misjudgement by the IBC over Jordan–Akins would bring their dreadful period as boxing's over-lords to an end. It was fitting it should happen so messily, a reflection of their own methods and a consequence of their refusal to believe they could ever be challenged, let alone caught.

The boxing world is not so easily shifted from its axis, though, and, while all this was going on, they arranged a return fight for Akins against Jordan, in St Louis in April. Jordan won again, and would go on to establish himself as a very fine welterweight indeed. He'd lose the title eventually, to Benny Paret in 1960, but he had come through the most tumultuous examination of his nerve away from his place of work. As well he might; Jordan was no stranger to the underworld, and mixed freely with Mob faces around Los Angeles.

History, meanwhile, suggested the IBC would get off lightly, just as they had in the past, just as many others had, crooked fighters, managers, promoters and gamblers. Not this time. This time, justice would be done.

The IBC was now dissolved under the anti-trust judgement set in stone four years earlier in the District Court. The rest of the castle would collapse with lovely symmetry.

The IBC guys had their day in court – a bad one, or several bad ones.

Carbo had gone down for two years already, on the face of it for acting without a manager's licence, and when he got out he had to face the music for the LA fiasco. First came a three-month court trial, the case against him brought by a young and eager Bobby

Kennedy, the US Attorney General, who successfully accused him of conspiracy and extortion and had him put away for twenty-five years.

He was also hauled before Kefauver's second Senate inquiry into organised crime and he performed no more convincingly than he had in many other inquisitions. He 'took the Fifth' as if programmed by a puppeteer, for once not being in charge of the strings himself, but responding to the instructions of his expensive lawyers. 'I cannot be compelled to be a witness against myself,' he parroted at least twenty-five times under heavy questioning.

Palermo, five feet of smiling menace, who had control at one point or another of some very fine boxers – Liston, Basilio, Ike Williams, Gavilan, Tony DeMarco and, almost, Jordan – went down too, but not so hard. Blinky got fifteen years.

Gibson? He pretty much got away with it. He was indignant that when the investigators moved in to cart them all away for questioning, he was picked up at a Chicago hotel, handcuffed and marched into custody, 'like a murderer'.

Gibson was no murderer. Not of lives, anyway. But he killed a few principles, butchered some dreams, injured quite a deal of trust put in him. He survived with a reprimand and a small fine. He regrouped, got out of boxing and went back to the sort of law he should have stayed with all along. Had he not been seduced by the bad guys, he would have had no part in this story. Had he not taken up Joe Louis's invitation to be his attorney, he never would have had to do business with the Mob. In the end, he did. He will say he was clean, that his field was law, that he advised, that he was an administrator of a business partnership and had nothing to do with the seamier side of the fight game. That would be as thin a defence as any proposed by other culpable parties in this story. They knew what was going on. By doing nothing, they connived in the spread of corruption, shakedowns, fixed fights and the general decline of an enterprise that resides so close to the edge of legitimacy even the slightest nudge away from propriety sends it into a tailspin.

That is what happened in California in 1958 and 1959. The wheels of the IBC came off and bosses were punished. What was left was an almighty power vacuum, a very dangerous situation in professional boxing. It would be filled soon enough.

There were many celebrations. *Time* magazine of 19 December 1960 observed:

> The supposition was that when the late Damon Runyon immortalized such citizens as Angie the Ox, the Lemon Drop Kid and Meyer Marmalade, he had largely consulted his own imagination. But last week, when Senator Estes Kefauver's antimonopoly subcommittee opened hearings in Washington on the fight racket, the characters who took the stand to describe the octopus grip of the underworld on US boxing were pure Runyon – but Runyon without romance.
>
> Dominating the proceedings from offstage was racketeer Frankie Carbo, 56, known to business colleagues as 'The Uncle,' 'The Southern Salesman,' 'Mr. Gray' and (in his younger, hungrier days) 'Jimmy the Wop.' Once convicted of manslaughter and five times arrested on suspicion of murder, Carbo is currently serving a two-year sentence for illegally operating as a boxing manager and matchmaker. In Carbo's absence, his pervasive influence over the boxing world was detailed by a man who should know: Truman K. Gibson Jr, 48, Negro ex-secretary of the now defunct ring monopoly, the International Boxing Club.

That was Gibson's lowest moment. He was derided now by the media establishment as no more than a stooge. They detailed his failings over a decade.

> In order to ensure that Carbo would make the boxers he controlled available for IBC fights, said Gibson, the IBC paid more than $40,000 to the ganglord's wife, whose last known address proved to be half a mile out in Florida's Biscayne Bay.
>
> There were other facts of life, too, Gibson admitted. The cartel paid $9,000 to Hoodlum Frank 'Blinky' Palermo, who is allegedly running Carbo's boxing empire while the boss is in jail. While Gibson doodled, Subcommittee Investigator John Bonomi summed up his testimony: 'Almost every leading manager or promoter in the U.S. is either closely associated with or controlled by Frankie Carbo in some degree.'

Time went on:

> At week's end, earnest Estes Kefauver, who is trying to decide whether a federal boxing commission is necessary to 'clean up the sport,' found another talkative witness: ailing (heart trouble) James Norris, millionaire ex-president of the International Boxing Club. Confirming most of Gibson's testimony, Norris added that Carbo had been a useful 'convincer' in lining up such boxers as Basilio and former Middleweight Champ Jake La Motta for IBC fights.

Dapper Jim had enjoyed his adventure in boxing, from the moment more than forty years earlier when he stood as a small boy alongside his dad and watched Jack Dempsey nearly murder Jess Willard. Now, this racetrack regular and round-town dandy wanted to leave Mr Kefauver and his stern committee with a smile to remember him by, and said that dealing with Mr Carbo had even 'embarrassed me with my horses'.

13

LONG LIVE THE KING

Some newspaper headlines – especially in the more serious broadsheets of the fifties when the media could be stiff as an ironing board – are so prosaic as to disguise the drama below them. 'KEFAUVER WANTS U.S. COMMISSIONER' was such a headline.

It sat above a single-column story in the *New York Times* of 30 March 1961, a plea for common sense, a reasonable call to decent people more than eleven years after the cat in the coonskin hat had reckoned, with all his heart, he could get rid of the wise guys. Now he wanted a commission to clean every corner of the stables in which he had waded knee-deep to 'get' the Mob. Certainly, he proved a lot of them had their fingerprints all over the boxing business. But, in the end, it was a bit of so-what for the fight game. A few went down. Some died. The rest moved on.

The guys had long gone from Jacobs Beach. The fight folk of the fifties were very much yesterday. There were new people moving in. They would run boxing by the same rules, always a couple of steps ahead of the good guys, and they would not lose sleep over what the story in the *Times* said that Thursday morning in the spring of 1961:

> WASHINGTON, March 29 (AP) – Senator Estes Kefauver, Democrat of Tennessee, introduced his controversial bill

today to place major professional boxing under tough Federal policing.

He told the Senate he was impelled in part by a fear that if the No. 1 contender, Sonny Liston, should win the heavyweight championship now held by Floyd Patterson, the title would 'revert to Mob control'.

Kefauver said there was 'compelling evidence . . . that Liston was controlled by three powerful racketeers' as recently as last December. He said he knew of no change since.

Kefauver's bill, as he had indicated, would create in the Justice Department the post of Federal Boxing Commissioner. This official would have the power to license boxers, managers, promoters and matchmakers.

Attorney General Robert F. Kennedy has indicated he doesn't like the idea of placing the commissioner in his department.

The bill would give the commissioner the rank and $20,000 annual pay of an assistant attorney general. He would have available the services of the Federal Bureau of Investigation and other Justice Department personnel to battle underworld infiltration of the sport.

Kefauver said he still hoped to win Kennedy Administration support for the bill.

He said in a statement filed in the Senate that hearings by his Senate Antitrust and Monopoly subcommittee had shown a need for Federal regulation to halt 'a massive conspiracy between racketeers and other undesirables'. He said that these men hoped to maintain a stranglehold on the promotion of big-time boxing.

Kefauver said that the hearings had produced evidence that Liston was controlled by 'three powerful racketeers – Frank (Blinky) Palermo, John Vitale and Frank Carbo –

operating through a front manager, Joseph (Pep) Barone'.

Kefauver said that the bill would give the Federal Government only 'simultaneous jurisdiction' over matches that figure in interstate commerce and would not pre-empt the field from state and local regulatory agencies.

The measures would require full disclosure to the commissioner of those who own or manage fighters, share in their earnings or have anything to do with promoting bouts.

Kefauver said it would enable an alert commissioner to detect secret deals and under-the-table pay-offs, such as those explored in his hearings. Violators would be subject to penalties of up to five years in prison and $5,000.

Few of Kefauver's noble objectives would see the light of day. Reality kicked in when the boxing community threw their newspapers away and got on with their deals. The match the wise guys wanted now was Liston–Patterson, whatever the establishment thought.

The Kennedy clan were not disposed to dirty their hands in prizefighting. Robert, the Attorney General, wanted the Mob crushed but his brother John, who outranked him somewhat, had more obvious reasons to lean towards pragmatism. The Mob hung out in every alleyway of American life. Including politics. Including Chicago politics, where Sam Giancana delivered the Democratic Party the votes they needed to put Kennedy in the White House, to give life to the new Camelot. So, JFK, the haloed hope of New America, stood back from the heat, while his righteous brother was tearing at the leash. Both would go down in the cause.

Anyway, to many in the FBI and other agencies of life control, the Mob was a beast they preferred to watch from a distance. It stirred now and again, and would be kept quite with a few well-placed kicks. But they knew, much better than the good senator, that what decent folk wanted was probably not achievable – not without a disproportionate amount of blood on some expensive carpets. Self-interest drowned out good intentions, not for the first or last time.

The New Order moved with the times. Soon, Cassius Clay, who won Olympic gold the year before, would well and truly open his Louisville lips and the fight game would roar again. His arrival, hardly lost on the leeches who cling to boxing, inspired renewed interest. At first, in what might have been a subliminal desire to be part of the mainstream, young Cassius attached himself to eleven businessmen from Louisville. 'In those first days after my return from Rome,' he would say later, 'I was proud to boast of my millionaire sponsors . . . I felt fortunate having so many people in town who wanted to give me what they called the "right kind of moral and ethical environment" for launching a career. As far back as I could remember, boxing was associated with stories of "gangster control", "fixed fights" and "back-door deals", some of it brought out by Kefauver Committee investigations while I was fighting in the Golden Gloves.'

Cassius was briefly in awe of his patrons. They had old Kentucky money, old Kentucky values and old Kentucky names. They were Southern royalty: Robert Worth Bingham, W. L. Lyons Brown Sr, Patrick Calhoun Jr, J. D. Stetson Coleman, Gordon B. Davidson, Archibald McG. Foster, George W. Norton IV, Vertner D. Smith Sr, Elbert Gary Sutcliffe, James Ross Todd and the main man, William Faversham Jr.

'I kept their names in my pocket,' Ali would tell his Nation-appointed biographer Richard Durham in 1976, 'ready to pull them out as proof of my status as a top-sponsored fighter, and I meant to display them as much as my [Olympic] gold medal.'

That was the same gold medal he was said to have thrown in the Ohio River when insulted by racists in a roadside cafe near a bridge in Louisville. As Jerry Izenberg, one of the most acerbic most a American writers on boxing, once cynically observed, 'If they trawled the Ohio for a thousand years, they'd find a mermaid before they found any gold medal.'

The gold would be tarnished soon enough, anyway. Clay quickly learned how to be as cynical as those who'd pulled his strings in the first place. The Louisville patricians did not last, and are seen through the prism of time as bloodsuckers. Phil Faversham, son of Bill Faversham (himself the son of an English actor), maintains the benefactors were 'people who already had money. It was an attempt to try and protect Ali, to keep the Mob away, to keep him out of trouble.'

With Carbo and Palermo banged up, Gibson de-fanged and Norris fast losing interest, there were still enough string-pullers to get Liston his shot at Patterson.

Floyd had been up and down during his reign and was there for the taking. Even at his peak, he would not have been able to hold off the fury of Liston. As he lost strength and speed, he resembled nothing so much as a deer wandering down to the waterhole at sunset, with the lion lurking in the long grass nearby.

When Liston finally got his hands on Patterson, in 1962, the papers all said it was a crying shame and John F. Kennedy, whose brother had just seen off two of the key mobsters in the history of prizefighting, was fully aware of Sonny's unsavoury dealings. JFK told the clean-living Floyd to whip this sonofabitch for America. Sonny knocked Floyd out in a round. Later, he did it again. Same round, same pain.

Sonny wasn't hot at the turnstiles, though, and, after being champion for seventeen months, the people he'd been naive enough to try to impress got him. Just as Liston had monstered good-guy Patterson twice, so Muhammad Ali, who would later become the most popular athlete in the history of sport, beat the quintessential bad guy twice, in '64 and '65. It was classic Americana. Each bout was suspect, especially the second, which lasted less than a round in front of nobody in Lewiston, Maine. The bets were in and, as Sonny's wife, Geraldine, admitted in 2001, Sonny was only ever up for one round of punishment. He knew he was supposed to lose, anyway. Same fee, less pain.

So who got to Ali? The Nation of Islam would arrive with a flash of righteous thunder. And Ali would be their all-time grandmaster of flash. Until he, too, palsied by the rigours of his work and consigned to the half-light of fading stardom, was sidelined in the end, an inconvenient reminder of the price the fight game extracts from even The Greatest.

The Mob had tired of Liston. He was never bankable, but he got them the title. He was extremely good at his job of creating immense physical discomfort and inspiring nauseous waves of trepidation in his opponents, which created its own aura. But he was also the most unpopular heavyweight champion since Jack Johnson. When he lost to Ali for a second time, in the backwater of Maine, with the world howling 'Fix!' there were, according to Truman Gibson, plenty of people at ringside who seemed decidedly unsurprised.

June 1966 turned out to be a busy month in the grand theatre of the fight game. In Texas, on the 28th, Ernie Terrell beat Doug Jones to keep the World Boxing Association version of the heavyweight title – not in itself a big deal, as everyone knew the real star was Ali. Muhammad had his WBA belt taken away for giving Liston that dodgy rematch in '65.

On 29 June, the day after Ernie beat Doug, Sonny was still working. In Stockholm, where they loved him – and where he and Geraldine met and adopted an orphan called Danielle – he took seven rounds to shred a German heavyweight, Gerhard Zec. That night, and well into the following day, Sonny celebrated . . . and several time zones away, back in the United States, a mother would soon give birth at the Cumberland Street Hospital in bleak Bedford-Stuyvesant, in no way the nicest part of Brooklyn.

Lorna Smith was born somewhere in the South in 1930, around the same time as Sonny. After World War II, and about the time Helen Liston was moving to St Louis, Lorna came to New York. She met and married a guy called Percel, of whom little was ever known. When Percel left, Lorna fell for Jimmy Kirkpatrick, a hard-living man with scant appreciation of his responsibilities – he had sixteen children by other women in the neighbourhood. He and Lorna gave life to Rodney in 1961, Denise three years later then, on 30 June in that World Cup summer of '66, a bundle of trouble they christened Michael Gerard.

Jimmy moved out before his second son had started to grow the teeth that would become infamous in a Las Vegas boxing ring many years later. So, although Lorna was known to neighbours in Bed-Sty as Jimmy's woman, she gave baby Mike the surname she'd held from her marriage to Percel, the name that was to become synonymous with cold, Liston-like terror: Tyson.

Sonny and Geraldine had moved to Vegas by 1966. He was still good enough to win fourteen of his last fifteen fights, but he made better money at his original profession: collecting debts for the Mob. For Christmas 1970, Geraldine went to her parents in St Louis with Danielle. When they got back to Vegas in the new year, the house stank, and Sonny's spent body was next to the bed, KO'd for good. Police found heroin and marijuana in the kitchen, morphine and codeine in his blood. Born in January, died in January, as his biographer Nick Tosches noted. One dead No. 1 Negro.

Lorna, meanwhile, struggled to support her family and had to move to Brownsville, an even poorer part of Brooklyn. Like Sonny, Mike was big but, lacking a father, he was short on confidence too. He had a high-pitched lisp and older kids called him 'fairy boy'; he was beaten unconscious four times. To escape, he went up on the roof of their building in Amboy Street and tended to his pet pigeons. He could hardly know then that, like Liston, he too would one day be the human equivalent of a pigeon, a pet of powerful people who cared little if he could fly.

When Mike was nine, a fifteen-year-old boy snapped the neck of one of his birds. Tyson flew into a rage and cleaned his clock, word of which reached some toughs called the Jolly Stompers. Armed with guns, knives and new bravado, he mugged his way into their dubious affections. He was a man at ten and, like Sonny, a mute thug by the time he was a teenager. Tyson and Liston never had the chance to embrace innocence, and will be judged unfairly for the rest of their lives because of it.

At thirteen, the age Sonny ran away from home, Mike was a serious delinquent and was delivered to Cus D'Amato, the man who, in the sixties, had kept Patterson away from Liston long enough to ensure that Floyd is remembered still as a world champion. By now Cus had semi-retired from mainstream boxing and lived in the Catskills in upstate New York, where his companion Camille Ewald watched him do his Burgess Meredith shtick. He shaped his scoundrel charges with growling homilies that might have been lifted from a John Sturges movie. And, getting old, Cus prayed for another champion.

His friend, the renowned boxing historian Jim Jacobs, knew they had a live one in Tyson. But he needed tutoring, so they showed him hundreds of old fight films, part of a priceless collection Jacobs owned with his partner Bill Cayton. Cus was particularly proud of Patterson and had all his fights on film. But Mike didn't fancy watching Floyd; he preferred the old guys, such as Jack Johnson and Jack Dempsey, Jack 'Kid' Berg and Abe Attell. And the fighter Cus feared like a Negro night – Sonny Liston.

What D'Amato could never crack was that there was a subliminal, unbreakable bond between Tyson and Liston, one bad black man talking over his wise white-haired head down the years to another bad black man. Sonny, who was fighting the very morning Mike was born, and died three weeks before the kid's fifth birthday. Sonny, it is said, didn't reach forty. And when Larry Holmes said years later that Tyson might not get there either, Mike didn't disagree.

When Mike was sixteen he went to the Junior Olympics and he knocked out everyone. The other boxers were heard to say, 'There goes Tyson. He's Liston's nephew.' D'Amato let the myth germinate. Some people still believe it.

That year, 1982, Mike's mother died of cancer. Two years later, D'Amato pulled some Catskills connections and adopted Tyson. There is evidence he also spent public funds on his prodigy which were intended for his community work. Cus knew the wrinkles. And he had his prejudices. 'Give him enough time,' he once said of his adopted ward, 'and the nigger will come out in him.'

When he died of pneumonia in 1985, D'Amato left behind not a young heavyweight with his head filled with wisdom (as is often written), but a spoilt and dangerous brat. Jacobs – another unusual man, who friends said made up stories about having a rich father and who fantasised he was Batman's sidekick Robin – ensured the early part of Tyson's pro career was a masterclass in salesmanship. But the nigger of Cus's imaginings was there for the taking.

Sonny had Blinky Palermo and a host of faceless bosses running his life. Muhammad Ali and Mike Tyson had Don King.

There would be no commissioner. There would be no more control of boxing in the sixties than there was when John L. Sullivan terrorised the bars of Boston eighty years previously. There would be cosmetic improvements. Fighters would be safer after being thrashed to within an inch of breathing. Occasionally, someone – usually Don King – would be hounded down and charged with some badness or other. And, nine times out of ten, it would come to nothing. He had seen the shenanigans of the fifties and learned his lessons. King resided as close to legitimacy as he could stand. At the time of writing, he still does. He was smarter than Carbo, smarter than Norris, smarter than nearly all of them. He didn't need to create trouble to prove how tough he was. He'd done with that in his youth, when he killed poor Sam Garrett.

But, just as La Motta epitomised one side of boxing, the grim pain and glory inside the ropes, so King stood for what went on outside the square.

King did refashion his life, no question. He desperately wanted to bury his past (who wouldn't with a rap sheet like his?), and he wanted to be taken seriously – but not too seriously. His wit has always been his secret weapon. When he completed what he likes to regard as his finishing school in the prison library, he went on to manipulate through the sheer power of his personality the most anarchic sport ever created. He has survived vendettas, at least two assassination attempts, serial vilification and institutionalised racism. He invented his own morality and his own strange stage act. His smile is pasted on but, in quieter moments, the facade slips and his eyes look as devoid of happiness as did those behind the hooded lids of Sonny Liston. King's resentment of prejudice and anger at the system – a system he has learned to play like a fiddle – is authentic. But nobody, least of all the man himself, is sure at what price.

Whatever the judgements moralists hang on him, King has insinuated himself into boxing's history like few others. His unique cackle remains the gravelled, drowning-out voice of the fight racket. He has gone from meagre roots to both help revive and disgrace boxing. He has put on the biggest and best fights, as well as some of the most chaotic and controversial, created the most noise and, by far, the most money – for himself and those who have hitched a ride on his wayward star. He has also bewildered and frustrated fighters, rivals and even his allies. Frank Warren, the London promoter who learned much from his association with

King, still admires the man, even though he lost a legal decision to him worth $12 million. 'He could have been anything he wanted to be,' Warren said once.

What King wanted to be, ultimately, was rich and powerful. He loves wealth, but he loves the influence it brings him even more. He moves in high circles. He has created his own kingdom, complete with logos, absurd hair and his special vernacular of outrageous gobbledegook. Nobody is more articulate in the language of hokum, exploitation and trumpeting the sometimes untrumpetable than The Don, the most outrageous, most entertaining fakir of our time and a life force the like of which none of us will see again. Few people in sport are hated, admired, feared, respected and fawned upon as Don King is. Nobody really 'gets' him. He is, alongside his creation Ali, the most interesting individual the fight game has thrown up in our time and, as hard as it will be for his legion of enemies to admit, he will be missed when he is gone. Like Eric Hebborn, the gifted painter whose copies of old masters were so good that art experts and auctions houses were left asking what was real, King's art raises the same question. King is boxing's ultimate forger. What he represents as real sometimes isn't – but it's usually good enough on the night.

Not many people get under King's skin; but the diminutive British promoter Frank Maloney got right up his nose. So frustratingly stubborn was the dyslexic south Londoner in deals involving the world heavyweight champion Lennox Lewis, King was reduced to calling him 'a pugilistic pygmy' and 'mental midget'. Maloney loved it. It made him a star. But, away from the megaphone, Maloney discovered a completely different King. 'We sat next to each other on a plane from London to New York once,' he recalls, 'and he didn't say a word to me the whole seven hours. As soon as we went through customs, though, his eyes lit up and he went straight into his routine. It was as if he were charging his batteries for the performance.'

I have two vivid personal recollections of Don King. The first was late into the evening of Saturday 24 September 1994, a few hours after his eccentric heavyweight Oliver 'Atomic Bull' McCall had thrown the best right hand of his turbulent career just thirty-one seconds into the second round, detaching Lennox Lewis from the world heavyweight title at Wembley Arena. We were in an annexe, mixing with a lot of excitable types who could hardly believe McCall, a crack addict since his early teens and a man of enormous mood swings, was the new title holder. King, somehow, had pulled another rabbit out of the hat – and now he was in full flow. I remember standing alongside him with Sri Kumar Sen, the boxing correspondent of *The Times*. We were smiling, laughing even, at the jubilant promoter's string of mangled jokes and anecdotes. Then, in an instant, King's face filled with thunder. A switch flicked somewhere inside that jangling brain and he let fly a river of invective at us about never having been given an even break, how white society hated his success, how he had to fight every day to beat down racism. I was seriously intimidated. So was Sri, by the look of it. We stood, frozen in silence, bewitched and shit-scared all at once. Was this how Sam Garrett felt when Donald the Kid laid into him outside his local bar in Cleveland twenty-eight years earlier? King ranted, uninterrupted. He had a chicken leg in one hand and a lackey stood beside him, holding a paper plate. When he'd worked his way through the meat, Don dropped the bones and the lackey made sure he caught them.

Then, like a dying monsoon, King calmed. The grin returned. His eyes went from menacing to sparkling. 'Heh! Heh! Heh!' I was not sure then and am even less sure now what had started the tirade or, indeed, brought it to a close. I do know I don't ever want to feel like that again.

Seven years later, one hot June night in New York, I met King again. He was in the middle of yet more litigation (he would win the case) and had invited me to spend a couple of days on his manor. It was rush hour in one of the world's most snarled-up cities and I asked him if he reckoned he could stop the traffic. 'You serious?' he said. He went up to a patrolman on traffic duty on Seventh Avenue and, as our photographer got down on his haunches in the middle of the road to record the bizarre event, King posed with breathtaking cheek, his teeth and eyes flashing as New York's traffic ground willingly to a halt.

'Only in America!' screamed King. 'Only in America!' replied the office workers and tourists crowded now on all available pavement space. King chuckled, to himself rather than with the audience. It was as if he was the King of New York. King of the world.

We moved on to a nearby hotel and, with the deftness of a nightclub bouncer, he got me past the goons of the Nation of Islam to meet Louis Farrakhan, that organisation's ailing leader. Farrakhan, born Eugene Walcott in the Bronx in 1933, had been a violin virtuoso as a schoolboy and a very good crooner in his twenties, a calypso singer who went by the name of the Charmer. He'd performed now and again in King's bar in Cleveland, when it was a raucous gathering place for inner-city black cool. The Charmer knew how to have a good time. Now, with his new name, new religion, new fervour, he wore the cloak of acquired saintliness. He eyed me cautiously, and spun a prepared tale of woe, about how he'd been banned from entering Britain for alleged anti-Semitic remarks. 'Tell them I want to come in peace,' he said. His beatific smile did not slip. Alongside him was a large, silent associate called Dennis. He was, I learned later, Dennis Muhammad and one of a few candidates waiting to take over the Nation when Farrakhan went to meet his maker, the original Muhammad. Dennis didn't exactly warm the room. King and Farrakhan went behind a screen and spoke for twenty minutes or so and then we rumbled on out past the Nation security guys and into the New York dusk. 'Got someone I think you'd like to meet,' King said as we walked the short distance to the Royal Rihga Hotel.

'Only in America, Don!' a passer-by shouted. 'That's right, my friend. That's right. Heh! Heh!' said Don.

When we reached the hotel, Dennis moved off on his own and we settled briefly at the bar. King hardly spoke as we sipped on iced drinks. Dennis returned and we took the lift up to the penthouse suite. We clambered out, and King came to life again.

When the hotel-room door opened, the man who filled it from inside, the hallway light giving him an eerie halo, was Muhammad Ali. The smile, mischievous, knowing and all-embracing, was so familiar I felt as if I'd known him all my life and we hadn't even been introduced.

King went into cheerleader mode, full blast. 'Muhammad Ali! Yes indeed. Muhammad Ali! And still the greatest of all times! My, my. My man, it is so good to see you . . . the greatest of all times! The history. The history.'

Ali stirred from his standing demi-sleep, clasped King to his chest and whispered, 'Yeah . . . and still a nigger.' Laughter filled the room. Even Dennis smiled.

Ali spread out on a sofa and started to drift, soft lamplight falling on his grey hair and a few fuzzy bits on his belly, exposed under a monogrammed shirt that splayed outside loose slacks. His wife Lonnie, a trained nurse who grew up on his Louisville street and had always adored him, fussed around his shambling presence. 'Want to change your shirt?' she asked. He shook his head, slowly.

It was Ali's turn to perform. He told a couple of stock jokes and went through his magic routine. All the while it was cruelly obvious he was fighting the fog of Parkinson's. He seemed to be having a fine time, nonetheless.

King and Ali reminisced. They go way back, to the sixties when Ali – Cassius Clay as was, and still a pup – went to King's Cleveland bar with one of his heroes, Lloyd Price. Lloyd is the guy who told King to straighten his hair and fix it into the shape of a crown, a hairdo to match his name. Sound advice. Lloyd had a couple of big hits back then: 'Personality' and, a favourite of King and Eugene Walcott, 'Stagger Lee'.

Ali looked at Dennis. 'I know you,' he mumbled, his eyelids drooping. 'From the old days, before you had your new name.' Dennis smiled weakly. King grinned. They knew each other all right. But Ali, weakened by his medication, couldn't pin it down.

King's cellphone rang. It was the actor Will Smith, who played Ali in the eponymous movie of the fighter. 'You gotta listen to Will, man! He sounds so like you!' Ali spoke quietly to Smith on the phone and the evening rolled on.

Then, suddenly and for no particular reason, King and Ali started humming bits of 'Stagger Lee', before collapsing in a giggling heap on the sofa. What, I wondered, was the joke? The story of the song, as I later worked out, bore an eerie similarity to that which unravelled on the April afternoon in Cleveland when Don King kicked a man to death. It might have been called 'The Ballad of Sam Garrett'.

Price's 1959 version tells the story of Stagger Lee, who goes to collect a gambling debt from Billy. The song ached with melancholy and was known in every ghetto. It had a life as varied as the characters within it, and it appealed deeply to King's devilish sense of humour. The story is widely believed to be that of Lee Shelton, a St Louis pimp, who murdered Billy Lyons on Christmas Eve, 1895, over a gambling debt, despite Billy's famous plea to be spared. When John Lomax immortalised him in song in 1910, it was not even established if the words referred to Shelton at all, but the myth was born. By the time Shelton had died of TB in 1912, he had become known up and down the Mississippi River variously as Stagger Lee, Stag-O-Lee, Stack-O-Lee and even Stacker Lee. His cruel and unforgiving heart was the acme of evil, and poor Billy was everybody's favoured victim, a man crying in vain for his life. Price, whose crossover rendering was so masterful it couldn't fail to be a hit, knew King's story too.

When Price appeared on national television soon after the song was released, Dick Clark, the nation's number-one arbiter of popular music and host of *Bandstand*, persuaded him to tone it down. No murders please, Lloyd, he said. Just a row. Price complied – and it still made number one.

Billy begged Lee to spare him. He would pay him if he could. Sam Garrett, life draining from his pulped body, had, according to witnesses, pleaded with King: 'Don, I'll pay you the money.'

Ali was getting real tired. Lonnie ushered us towards the door. King, arm around the man who'd made him famous, whispered something to him as we left. Ali looked up, smiled one more smile, and repeated, 'Yeah . . . and still a nigger!'

We climbed into King's limo and he slipped his chauffeur a hundred dollars. Neither blinked. It must have been the regular tariff for any journey, and there could be several on any given day or night. We drove no more than a few blocks – and could easily have walked – to one of his favourite restaurants, Ruth's Chris Steak House on West 51st Street.

I turned off the tape recorder, and we relaxed. King wound down, spoke more quietly, gently almost. He was a little tired, which is a rarity in itself. King is known for his almost inexhaustible energy and will regularly wear down business rivals, staying awake for days before delivering the *coup de grâce* when they are wrung out and on the ropes.

Now, his eyes dimmed and his big shoulders slumped. King invited me to order the wine. 'Anything you like, my little Limey friend,' he said. I baulked at the top end of the range – $1,000 and more – and settled for a couple of $40 bottles of Napa Valley red, sure that he would stiff me for the wine bill at the end of the evening. He didn't.

As the night wore on, way past midnight (they'd kept the restaurant open for us and we were the only customers left), King opened up. Was he happy? I wondered. He'd wanted to be a lawyer, he said. All he ever wanted was respect. He knew the white establishment would never give it to him freely, he said, so he went out and took it. Don moved in his own way among his own people. He made things happen. He used the system. He explained the numbers racket to me, wanted to know how it was in any way morally different from the Irish Sweepstake, just a little untaxed fiddle after all, a way for poor folk to have a shot at some money. He knew Colin Powell, America's soldiering hero in the first invasion of Iraq, and, at the time, courted by the Republicans as a presidential hopeful. 'His mother', King whispered, 'played [the numbers]. She hit one day and got enough money to move them out of the ghetto and Colin ended up in a nice neighbourhood and went to a good school. So, what you want to say about that?'

He told me he read Mao and Marx. He met Castro in Harlem. 'He was born and bred in the struggle,' King said of the Cuban who Americans are encouraged to despise above all others. 'While we've got the propaganda machine that has been critical of him, he's been in office close to forty years. You don't stay in office in the Third World and emerging countries that long. So it can't be true what they all say about him, because the people still support him. Propaganda is a lethal weapon, man. A lethal weapon. It makes you believe things you never would believe, and you know the reality of it is different.' He said all this without missing a beat, without a smile to betray the irony of it all.

As we got up to leave, full of a fine meal and some not-bad wine, he pulled a sizeable wad of notes from the inside pocket of his silk-lined jacket and began to peel them off in the most casual way. $500 each note. There were, I think, four on the table as we left. Don did not look back, nor did he ask for a receipt. Some tip for a bill of perhaps no more than a few hundred dollars. The staff beamed as if they'd been handed the heavyweight championship of the world – and don't think that prize has not been in Don King's gift from time to time.

Outside, I leaned down to the window of the limo and said goodnight to the chauffeur, who was getting ready for his second one-spot in the space of an hour and a half. 'Does he always tip like that?' I asked. 'No,' he said. 'Sometimes it's more.' King got in and waved me to follow. 'Come on, my little Limey friend. The night is young!' It was getting on for three o'clock in the morning. I was Kinged-out. 'Thanks, Don. I'll see you in the morning.' Maybe I should have hung in there. King was on his way to another part of his manor. Before dawn, he would sit down in Harlem with a man who spoke his language far better than I did: Mike Tyson.

I wasn't about to believe everything Don King told me. But I could see how a lot of people would. I think I understood, finally, why he survived. I knew now why he made boxing his jungle. It was the only place he had respect. It was the one place that would have him without reservations. Apart from the streets and prison. He was no saint. Far from it. But, in boxing, he didn't have to be. In boxing, he could tell any far-fetched story he wanted and it would be good enough to pass for the truth. King didn't invent bullshit; he just made it smell pretty.

Don would shape boxing the way he chose. And others would fall in behind him, because that's where the money was. There are no morals in the boxing business. The rule that matters to the uncomplaining majority of people in the industry is this: make as much as you can and keep your mouth shut.

14

TEDDY ATLAS KNOWS

Teddy Atlas is no gangster. But he's as smart as one. And he has survived in the boxing business by keeping a sense of perspective, which might be another way of saying he doesn't make too many value judgements about people he has to do business with.

Atlas was a good amateur boxer, and became a great trainer. He put the shine on Mike Tyson when he was a teenager, only to be dumped. Unfazed, Atlas went on to train Michael Moorer and, memorably, sat on the fighter's stool between rounds to prod him into life against Evander Holyfield. Moorer, a talented but sometimes immovable lump of a heavyweight, responded by going on to win the fight and the world title. So Atlas, clearly, has the gift.

The son of a doctor, he ran wild in the streets as a kid, which is how he ended up knowing Tyson, at Cus D'Amato's little hideaway for scallywags in the Catskills, in upstate New York. Today, he is an impressive ringside commentator.

Yes, he agrees, the fifties were corrupt. But that's boxing, whenever and wherever, is how he sees it. I get the impression Atlas learned his pragmatism at the feet of a master in D'Amato.

'Cus was part of the Managers' Guild, yes, but he fought the IBC and he thought that in the end he was in some way a part of their downfall. He felt the IBC had a monopoly, and a corrupt monopoly at that, like the Garden which was owned by the IBC and Jim Norris and all those guys, or was partners with the IBC or whatever way you want to look at it, and Cus thought that it was no better than today.

'It was a corrupt organisation. Maybe they kept things in line better; but there were fewer champions and less confusion and in that way I guess there was some semblance of control and more of a feel of stabilisation knowing that there wasn't a million champions and a million organisations. There was only one bad organisation instead of four or five. Some people may not agree but that may be the fairest way to actually call it.

'The IBC was shown to be a monopoly of not always fair proportions and Cus was fighting them. Cus's goal was to get the heavyweight champ Floyd Patterson and then once he did he was to announce on that day when he won the title that he would not defend against an IBC-controlled fighter. Then later on you had the Kefauver hearings and the IBC went away. And Carbo and Palermo, although it took a little longer, eventually went away.

'I can understand them saying [it was better then] but it doesn't mean to say there weren't uneven things going on and unfair things going on – and more accurately and to the point corrupt. It's also not fair to say it was just the fifties. It went way back. But if you talk about a golden era of fighters, the forties, the thirties, the twenties, you didn't have that many fighters who weren't fairly capable. Some were better than others but all of them were fairly capable; now you have many who are not capable. I used to get so upset on ESPN when Max [Kellerman] used to say that Roy Jones was the greatest fighter of all time. I wanted to break the damn monitor and I would say, "Max, my God, how could you say a thing like that? Forget about the books you've read, you've got to go beyond that to something tangible and things that matter." And what matters is what these guys did and what kind of fighters they fought. He [Jones] couldn't handle Archie Moore, he couldn't handle Harold Johnson, and then some guys whose names we didn't know so well. There were a lot of them that he probably couldn't handle.'

Atlas is also a gatekeeper of D'Amato's legacy and, while conceding he was far from perfect, he reckons his philosophy and behaviour were just right for the environment in which he found himself. He did not agree, for instance, that D'Amato was over-protective of Floyd Patterson, a point most people would regard as proven when the champion was eventually fed to a mobbed-up Sonny Liston.

'But that's silly,' says Atlas. 'Floyd fought a lot of good fighters. He fought a lot of good fighters before he beat Archie Moore for the heavyweight title. Archie Moore was an old guy but Floyd beat a lot of good fighters. Cus brought Floyd up and turned him into the heavyweight champion of the world. Sure, he took advantage of the landscape, he used management skills to navigate and manipulate; there is no doubt about it and there is no doubt that Floyd got knocked out by Sonny Liston but there was no doubt also that Cus developed a small guy who had good hand speed, who developed a certain style that Cus felt was a style that suited him.'

And what of the criticism that D'Amato was not so much an angelic philanthropist but a greedy man, just like anyone else negotiating the shark-infested waters of the boxing business in the fifties? Because Patterson, really, was no heavyweight; it was just that D'Amato knew that's where the big money was.

'Again that's ignorance. It's people looking at the icing on the cake. Cus was motivated by money. The heavyweight was the thing that was most spectacular, Cus liked things that were spectacular and people were going to notice the heavyweight champ more than they were going to notice the light-heavyweight champ. Cus had other agendas and other purposes that were attached to his goals and one of them was wanting to have some control over boxing. He could do that better if he had the heavyweight champion of the world, especially in those days: if you controlled the heavyweight champ, you had clout. And for him to be able to do some of things he did later on, like battling the IBC, those things would have been much more difficult if not impossible if Cus would have only had the light-heavyweight champion.'

Not that Atlas thought D'Amato was unblemished. Far from it. But he is circumspect about some details. What was his take, for instance, on the renowned trainer Ray Arcel, who was hit over the head with a lead pipe for daring to challenge the gangsters?

'There are stories that I can't get into, that I can't talk about. About certain affiliations. And anyway, I don't know for sure because I wasn't there. I would just be relaying stories told to me and it would only be one side of the story. But, look, Cus wasn't exactly Father Flanagan either. He said that he fought the IBC and he was the only one to do it but again it's the icing on the cake. What's inside the cake? I mean the perception sometimes becomes more reality than the actual reality.

'Cus didn't have certain problems although he fought them [the IBC]. Now, you have common sense, so you ask, how come he didn't have problems? It's common sense that if Cus fought these people – we don't know exactly what happened with Arcel – and nothing happened then there must have been something. Or maybe they just liked people with white hair cos Cus had white hair and he wore a bowler hat.'

Slowly, Atlas lifts the lid on the past.

'Later on, when I left Cus, some of those reasons were maybe brought out a little. Cus had always talked about a guy he was friends with called Charlie Black. I found out that Charlie Black's real name was Charlie Antonucci and Charlie Antonucci was cousins with "Fat" Tony Salerno. Tony Salerno was the head of one of the biggest Mob families in the country so Cus being a guy that was wearing a collar and fighting this good fight and somehow mysteriously the rain never fell on him . . . suddenly it's maybe a tiny bit explainable that he was with Charlie Black.

'Two and two is usually four and I don't have to be Nostradamus or Columbo to say that is maybe why he wasn't bothered. Maybe that upsets a few people that wanted to have a different remembrance, but who created that history? I'm just stating something that is a solid fact. I'm not drawing a conclusion. Listen, I'm a guy who spent seven years with Cus and have a lot of kind thoughts and a lot of grateful thoughts about that time but I also have other thoughts too, about some of the manoeuvres and some of the things that Cus did. I was disappointed in him. Disappointed in the end. I developed Tyson, did most of the work with him, and then all of a sudden this one fighter brought about a compromise in some of the things that Cus had led me to believe meant something.'

Tyson was D'Amato's masterpiece. In a way, it was a shame he didn't see him win the title; in another way, his death saved him from witnessing the fighter's decline.

'I think he saw him more as Liston. To get revenge for the guy who obliterated his guy [Patterson]. I think he saw him more like Liston than Marciano. Then you don't have to change the colour.'

Atlas challenges all sorts of perceptions, across the board. He does not, for instance, view the fifties as a golden age of boxing.

'I don't think it peaked then. I think we think it peaked because that was the last glimpse we had of it, but it was steady and at its best in the thirties and forties. It began to slide a little bit in the fifties. Maybe you think it peaked because there was more places to see some of it. Coverage had progressed and you didn't have fights that weren't able to get to the public eye, and there were less bad decisions.'

Towards the end of 2007, Bill Gallo, an old-timer with a long memory, wrote in the *New York Daily News* about how he had no time for pessimists who keep reading the last rites over the corpse of boxing.

'Well folks, boxing is not dead, but neither will it ever be king again,' he wrote. 'The most it could ever be is the jack of clubs, which is a helluva lot better than the deuce boxing was holding for the last eight years or so.'

Gallo was enthused by the super-middleweight contest between Joe Calzaghe and Mikkel Kessler, which drew more than 50,000 to Cardiff's Millennium Stadium in the middle of the night. He was also roused by the prospect of another night in the Garden, to watch the coming Puerto Rican Miguel Cotto against the Californian Sugar Shane Mosley, whom Eddie Futch once described as potentially the equal of Sugar Ray Robinson. Gallo was not disappointed. Cotto beat Mosley in a classic. Some nights, the rough old game still gets up off its knees to surprise us all.

15

SPARRING WITH SCHULBERG

As long as he lives – and, with the best will in the world and all the angels singing for him, it can't be that much longer – Budd Schulberg will never write a finer line than the one he wrote for Marlon Brando in *On the Waterfront.*

'I coulda been a contender' doesn't just hold up in the context of the film's wonderful narrative, there is simply no better evocation of the frustration and addiction of boxing. It might spring from the soul of the most ordinary pug, or from a gifted artisan of the ring who just didn't get the breaks. It is a universal plea. And the little man who wrote it more than half a century ago knows where it came from: the docks of New Jersey.

Schulberg has spent all his life – more than ninety years at the time of writing – around the fight game. And, in researching the story that would go on to win an Oscar for his script, he listened long and hard to the argot of the docks, where fighters and mobsters mixed, where lives were made and broken.

In the summer of '07, I went to see him for the *Observer*. It was a journey well spent . . .

To get to Budd Schulberg, you have to go through Hicksville and Babylon. They are stations on the Long Island Line from Penn Station to his home in the quiet hamlet of Quiogue on the edge of Westhampton, a chic little celebrity hideaway on the Atlantic coast two hours out of the Manhattan he loves. But Hicksville and Babylon could easily pass for metaphors in the long life of one of the twentieth century's most endearing literary survivors.

Schulberg has seen his share of small-town hicks and shysters, in and around boxing rings – and he's been a witness to the high life too. There aren't many major names in show business and American sport over the past eighty years or so that he hasn't met or known. He has much to reflect on and will do so again with his pen when he gets round to writing his autobiography.

Meanwhile, looking out over the quiet creek at the back of his rambling, comfortable bungalow, keeping a check on his medication and taking advice on all manner of things from his fourth wife, Betsy, Schulberg was planning his next adventure.

He'd received a phone call from London that would disturb what laughingly passes for his retirement. 'The director Steven Berkoff was on the phone,' he said through the stutter that has been with him since childhood, 'and he wants to do it in the UK.'

'It' is *On the Waterfront*, one of the landmark movies of our times, and which Berkoff fancied doing as a stage play. It was no given it would work – a musical version ran for only eight performances on Broadway in 1995, but it is typical of Schulberg's innate pugnacity that he could not wait to try.

Schulberg wrote the script in 1954 and gave Brando perhaps his finest hundred minutes on the screen. Contained therein is the distillation of the boxing industry. It is worth recalling in full the words Schulberg put in the pouting mouth of Brando's Terry Molloy, the washed-up pug reduced to strong-arm work for his brother Charley, whom he berates in the back of a taxi one night.

'It wasn't him, Charley!' he tells him, pleading that the blame for his failed boxing career lay closer to home than an unnamed wise guy who'd been pulling the strings offstage. 'It was you. You remember that night in the Garden, you came down to my dressing room and said, "Kid, this ain't your night. We're going for the price on Wilson." You remember that? "This ain't your night!" My night! I coulda taken Wilson apart! So what happens? He gets the title shot outdoors in the ball park – and what do I get? A one-way ticket to Palookaville . . . I coulda had class. I coulda been a contender. I coulda been somebody, instead of a bum, which is what I am . . .'

Where, I ask, did he conjure that up from, that one sentence – 'I coulda been a contender' – wrapped inside an angry rant? He didn't think the words were that special when he wrote them.

'I was so immersed in it. I don't think I stood back from it and thought that. I had hung around these guys an awful lot by this time and I'd sat up with them, drank with them at the bars, sat up with them in their cold-water flats. I really had taken a lot of time to know them. I picked up on their language. After a while I didn't have to stop and think about it. The words came to me naturally.'

Is it Brando's longest speech in the movie?

'I think it may be, yeah.'

He's largely inarticulate for the rest of the movie – in words at least – but he gets very emotional with Charley and lets it all out. You see boxers who've been hard done by and, often, they don't know how to express it. Terry Molloy did that for them, didn't he? Yet there's a sense of powerlessness in that speech.

'Yes. I think so. It's all inside of them. They're angry, but they can't verbalise it.'

The movie – and Schulberg's part in it – was mired in controversy at the time, described by the left as an unsubtle attack on the union movement, and Schulberg was also infamously caught up in the House Un-American Activities Committee investigation into communism in Hollywood. As a former party member, he was widely derided for naming names and wrecking the careers of former members and colleagues.

Whatever the rights and wrongs of his stance, Schulberg has consistently defended his actions ever since. As he saw it, he was hitting back at what he regarded as the attempted censorship by the party of his first novel, *What Makes Sammy Run?*

That 1941 book, ironically, put him on the outer edge of the film industry for many years, partly because John Wayne, whose clout was hardly negligible, considered it 'a Communist plot' as it advocated the formation of a writers' guild. Schulberg's father, the renowned Paramount studio boss of the thirties, B. P. Schulberg, had himself fallen off the greasy pole. So Budd knew a thing or two about being ostracised by the time he made one of the great comebacks with *On the Waterfront*.

He said once, 'I don't feel what some people expect me to feel. What's painful is to have believed in something that sounded so right, and that turned out the way the Soviet Union turned out. It's more the disillusionment that hurts for me.'

What is often forgotten is that, during his testimony to the HUAC in 1951, he defended many party members as 'idealistic' and 'innocents'. He made a case for them to be exonerated because 'they got into something that they really didn't understand'. But Senator Joe McCarthy would not be swayed from his mission and a lot of those innocents were metaphorically slaughtered. Some of Schulberg's associates never forgave him for his part in the whole sorry business.

Schulberg, although badly bruised by the experience in those most turbulent of times, soaked up the punches well enough.

He saw *On the Waterfront* as the perfect vehicle, a story with which to expose the influence the Mob were having on decent dock workers, men who had no alternative but to bow to the bullying. And, of course, he saw it as a way back from the literary wilderness.

'I spent several years hanging around on the waterfront, working through that waterfront with Father John Corridan [renamed Father Barry and played by Karl Malden in the movie]. He was fascinating, one of the most fearless men I met. Ever.

Holding meetings in the bottom of the church, very much the way you saw it in the movie, with the rebel longshoremen. He was acting, in a way, like a labour organiser. He was organising these men, and guiding them in standing up to the waterfront racketeers, the men controlling the union, the ILA [the International Longshoremen's Association].

'The Mob really ran the damn thing. The ILA and the Mob were interchangeable. I based the movie very much on Local Eight 24, which was called, for very good reason, the Pistol Local. These guys literally shot their way in. They just hijacked the union. And they ran it with a brutal, brutal hand. Once they controlled the hiring bosses, the guys, who just as you saw very much in the movie, picked them and said, "OK, you and you." There were kickbacks.'

Did he see, I wondered, a parallel here with boxing, where Frankie Carbo, for instance, would say to Jim Norris, or whoever, 'I want a piece of that fight"?

'Very true. That's the way it was. It was. It was! Total intimidation. Total. Every so often [fighters] were told, "This is not your night."'

It sounded as if the Mafia translated the thuggery of their street operations directly to the ring, I said, employing prototype Mob method for controlling people's lives.

'Yes, it was like that. Pure intimidation. Absolutely, total. And very brave men became manipulated by these men. I don't think there's any braver thing to do in sport than get in a ring with another man and put everything on the line, and yet those men were intimidated by the Mob.'

On the Waterfront hit a nerve in American life that few had addressed so eloquently. Then Schulberg went a step further. He wrote about the gangsters and their involvement in boxing for *Sports Illustrated*.

'I wasn't so afraid somehow in the magazine but then they took out a full page ad in the *New York Herald Tribune*, full page with my face all blown up. When I saw that I felt a twinge of fear, I did. Nothing happened, though – except they barred me from going to the Garden.'

Schulberg knew the principal actors in the drama – Carbo, Palermo, Norris and Gibson – and didn't think much of them as a group or human beings.

'Truman knew how to deal with power. He really was very smart, shrewd, but he was not . . . he had no real . . . qualities. The Mob were very easy to get on with, really, and they wouldn't stop him doing something if they thought it was advantageous for the business.'

As is the way in the film industry, there was drama away from the set even before the cameras started to roll. In his original draft, he'd wanted Terry Molloy to die at the end of the movie. This brought Schulberg into conflict with one of the most powerful men in Hollywood.

'Well, it was a big argument, I can tell you. On the last day of the filming, the very last day – in fact we shot the final scene on the last day – I was still saying I thought he should get killed. I thought it should end dramatically. Sam Spiegel, whose money was involved, he thought it would be a disaster if we killed him. Elia Kazan [the director] was kind of in the middle, the way he often was with us. Usually the argument would be between Sam Spiegel and me. With "Gadge" [Kazan's nickname, short for Gadget] sort of the swing vote. I told him it would be very effective if he had that in, so that's why I did that in the book.'

In the end, as ever, money drowned out debate and Molloy lived. What Schulberg also told me, though, was that Brando came close to missing out on the role that defined that stage of his career. The role could have gone to Frank Sinatra, whose own career needed major surgery in the early fifties and who, coincidentally, was hanging around with just the sort of wise guys who would make Terry Molloy's life a misery in the movie. He even came from Hoboken, where the movie was shot.

'It was complicated. We went after Marlon first, and he turned it down. Then we went to Frankie and he loved the idea. It was perfect for him. Then Sam Spiegel – the thing to be a great producer is to be a great seducer – he wined Brando and dined him. Sam kept saying, "This is Brando's. Marlon should be doing it." Then somehow, by God, after we got poor Frank on board – they were even doing wardrobe on him – he brought Brando back in it again. Frank was infuriated. I can still hear him screaming at Spiegel. He sued in fact. They settled.'

Terry Molloy was based, ever so loosely, on a real-life boxer, Anthony 'Tony Mike' de Vincenzo, which, at first, didn't greatly please him. A year after the movie came out, de Vincenzo won a small settlement, out of court, alleging the story was too close to his own life to be coincidental.

Schulberg maintains it was largely a coincidence – but concedes there are similarities.

'I think I had the character in mind pretty much before I met Tony,' he says, 'but Tony Mike was a decent boxer also – so everything about Terry Molloy, in a sense, overlapped with the actual experiences of Tony Mike. But I had not interacted with him so much. What happened, really, was that it was Marlon who knew him very well. Marlon bonded with him and I didn't stop it because I thought it was really good for Marlon. And so the identification with Tony Mike, I would say, became more strongly identified than I felt at the beginning. I knew that character, that type rather than just only Tony Mike.'

Was Brando into boxing?

'No. The thing about Marlon was he was the damnedest natural I've ever known. I asked a friend of mine to teach him and take him up to Stillman's gym and box with him. He came back after the first day and said, "Jesus, Budd, this kid is a natural. I can make a hell of a fighter out of this kid." He really meant it. He said he learned so fast. "I jab with him, he does it right away and he does it so well." Marlon, he was a strange genius, who could do anything. He could do the bongos; if he wanted to be a dancer he could dance. He could do anything.'

And the irony is, he didn't have to box in the movie – just swing in a street fight.

For Schulberg, those were the days of greatest excitement. He'd seen off the doubters. He'd survived the trauma of the McCarthy witch-hunt. And he'd written one of the great scripts.

After *On the Waterfront*, Hollywood dipped into Schulberg's archive again in 1956 and put his 1947 novel, *The Harder They Fall*, on the screen. It has been rated among the best boxing films of all time. In the pre-release version for selected critics, Humphrey Bogart, the disillusioned writer, is seen typing, 'Boxing must be abolished in America.' Schulberg insisted – successfully – that this be changed to: 'The boxing business must rid itself of evil influence, even if it takes an act of Congress to do so.' Estes Kefauver was trying to do just that.

Budd still gets to most of the big fights, accompanied by his son, Benn, and writes a regular column for a Scottish newspaper. The obvious question as he approaches his century is: why?

'Well, I'd call it a very deep habit, really. I've been attending boxing matches since I was, maybe, twelve years old. My father was a big fan and, religiously, we went twice a week, Mondays downtown at the [Los Angeles] Olympic and Friday nights at the Hollywood Legion. So I fell in love with it that way – just the way Benn has.

'You can see why boxing appeals to so many, many writers: Conan Doyle and Bernard Shaw, up to Jack London and Hemingway and Nelson Algren and all the rest. The reason is a very simple one: boxing is the most dramatic, one-on-one of all sports, more than say . . . tennis would be, because of the sense of danger that's involved, that's one thing, and the intense feeling of two men attacking each other.'

With Schulberg, it always comes back to the words. They are his jabs and hooks. After the success of the movie, Schulberg wrote the book version. He'd picked himself up off the canvas. He'd beaten the count of Hollywood, an industry he despised yet depended upon.

'We had an era in boxing literature, going back some way now, where there were lots of good writers around. Dan Parker, Reg Smith, Jimmy Cannon, all those. That was a golden era. Damon Runyon was another. There was a sense of drama for those boxing writers that you didn't see in any other part of the newspaper. Pretty often the best writing in the whole paper was there in the sports pages. Jimmy Cannon called boxing the red-light district of sports and it's always had that seedy side to it. At the same time there's something noble and even heroic about it.'

Nevertheless, Schulberg has reservations about how some writers were drawn too deeply into the business, how they looked the other way. In 1950, he described Mike Jacobs, the grand lord of the Garden for two decades, as the 'Machiavelli of Eighth Avenue'. He observes now of the curious Jacobs, 'He didn't seem to have any personal friends at all, yet everybody called him Mike. People played up to him. But I don't think I remember any of us who could say he was a real friend. He was strictly business all the way. Totally unsentimental. And he had all that power.'

Schulberg has been around long enough to see most of them come and go. He might not have missed Jacobs much, but he was hardly keen to see him replaced by Norris, Carbo and Palermo. The other enemy, he remains convinced after all these years, was television.

'What TV did was make a few big stars at the expense of the good, run-of-the-mill fighter. In those days there was always a fight somewhere, every week, in New York City, practically every night.

Those days are gone. You had to be good just to get into the Garden on the undercard.'

Budd got there. Once. As a manager. Only old-timers will remember that the writer who now rails against managers, promoters and the evils of television was once briefly a manager himself.

'My heavyweight, Archie McBride, I got him a semi-final in the Garden. We were really all excited about that. That was a big deal.'

McBride was a willing fighter not quite good enough to break into the upper echelons of the division, but he and Schulberg had fun. When Archie's regular trainer, Whitey Bimstein, was unavailable, Budd even sparred with him before one fight – not quite as physically as he might have done but hopefully with more vigour than he showed against Ernest Hemingway. The Old Man grabbed a young Schulberg by the lapels at a party in the thirties and rammed him up against a wall, demanding he proclaim his boxing CV, almost baiting the small, anaemic Schulberg into fighting him. Budd demurred and, not for the first or last time, Hemingway made a fool of himself in trying to assert his machismo and fighting prowess.

Schulberg's involvement in pugilism was an innocent diversion by comparison. He recalls of their foray into the dark side of the business, 'Archie made enough from it to buy a little house and enough to put his son through college, and that was the best thing about it.'

Archie McBride sounded like a cartoon boxer, a Joe Palooka or Canvas-Back Harry – but he was a reliable operator. He boxed for twenty years and, while never threatening to win a world title, he did fine. McBride came up from Brewton, Alabama, in the forties and made a home in Trenton, New Jersey. He won twenty-four of his forty-six fights, stopping just ten opponents and getting stopped six times himself. It was a classic middling career in the fight business. McBride fought at the Garden three times, knocking out

Lloyd Gibson in 1951, losing on points to Tommy 'Hurricane' Jackson two years later, and going seven rounds with Floyd Patterson in 1955 before Floyd knocked him out.

Thereafter it was back to St Nick's and off to Germany and all points east of Manhattan. He was a journeyman fighting man, but proud and willing. He had his moments. He beat the hard-hitting Bob Satterfield, lost on points to Ingemar Johansson in Gothenberg and was decisioned by the fine Cuban Nino Valdes in Havana. Towards the end, in 1965, Doug Jones KO'd Archie in five – but then he nearly did the same to a rising Cassius Clay two years earlier in the *Ring*'s fight of the year.

No, Archie did as well as he could have imagined when he started out in 1947. And Budd was happy to be along for part of the journey.

'It's so intensely personal, boxing,' Schulberg said. 'There's nothing really like it. In boxing, one night can change your life. That's what you get in boxing. You don't play boxing. It's deadly.'

16

ROCKY DIED IN ONE OF MY JACKETS

Al Certo hung in and around the Garden most of his life. Not many know the workings of the fight game better than the tailor from Secaucus.

'I was born and raised in Hoboken, New Jersey,' says Certo. 'Where Frank Sinatra was from. In fact I knew his father, Marty. And I knew his mother, Dolly.'

Certo has been near the heartbeat of boxing for most of his fourscore years. He grew up alongside legends of all sorts, good and bad, mysterious and infamous, and has a thousand stories, some of which he remembers better than others, some of which, for his own peace of mind, are probably best left undisturbed.

He's been up there. He's trained a world champion. In 1991, the Boxing Writers Association of America voted him their manager of the year. Three years before that, his citation when he was inducted into the New Jersey Boxing Hall of Fame read:

> Al Certo, who boxed under the name of Al Certisimo, was born in Hoboken on September 28, 1928. At the age of 19 he took up amateur boxing and capped an unbeaten amateur career by winning the Golden Gloves title in the 135lb class. Certo turned pro and the winning streak continued. He reeled off 20 triumphs, including 11 kayos,

> before losing to Benny Hernandez in Brooklyn. [Certo's pro career lasted ten days short of a year, between 26 March 1953, and 16 March 1954, and his official record reads: nine wins (four KOs), one loss – so maybe the eulogy took into account his amateur CV. Nevertheless, Mr Certisimo must have been pretty good.] Certo's career was cut short when he cut his hand with a saw while helping his brother remodel a car wash. Certo didn't stop there. He has run many benefits and fund-raisers, bringing in such greats as Rocky Graziano, Willie Pep, Jake La Motta, Joey Giardello, Tippy Larkin, Carmen Basilio, and Jack Dempsey just to name a few.
>
> Certo later became an outstanding promoter and manager, handling some of the top ringmen of our day including former world junior welterweight champion Buddy McGirt. It is with great honor that we induct Al Certo into the New Jersey Boxing Hall of Fame.

Applause all round.

Five years later, in 1993, the glowing New Jersey perspective on Al Certisimo of Hoboken was not wholly endorsed in the *Philadelphia Inquirer*. The old-fashioned multi-decked headline of that newspaper ran thus:

TESTIMONY TIES BOXING MANAGERS TO MOB
THE TWO MEN RUN BUDDY McGIRT'S CAREER
ONE DENIED THE CHARGE
ANOTHER TOOK THE FIFTH

So, what happened?

On April Fool's Day, 1993, Salvatore Sammy 'The Bull' Gravano, one of the Mob's greatest 'singers', testified before the Senate Permanent Subcommittee on Investigations that the co-managers of McGirt, Certo and Stuart Weiner, were 'associates' of the New York's Gambino crime family. This is a conveniently imprecise description, one that invites speculation across the spectrum of doubt.

Certo, just about bursting a blood vessel in his sixty-four-year-old guinea neck, said it was a lie; Weiner pleaded the Fifth Amendment. You could call it a split decision.

Gravano, who'd admitted to being party to the killing of nineteen people in his time – but pulling the trigger only once – had moved assuredly in the underworld all his life. Then he ratted on John Gotti, the Gambinos' main man, for murder and racketeering in 1992. From that point on, he was despised on all fronts.

Gravano told a journalist once, 'Sometimes I question how I got to be the way I am. I had a great mother and father. I don't know how I got to be the way I am. No emotion. No feeling. Like fuckin' ice.' You never knew where you stood with Gravano: victim or cynical manipulator. He lived by the oath of Cosa Nostra – and he lived a little longer by breaking it.

They called Gotti 'The Last Don'. Which might have been wishful thinking – there is always a frog waiting to turn into a prince. Like Al Capone, to whom he has been compared, and Frankie Carbo, Gotti dressed pretty and thought ugly. All had fierce tempers. All were killers of the top rank. But, unlike Mr Gray, Gotti loved his name in the papers. It was as if he wanted to be wiretapped. When the FBI went after him they had a library full of footage, miles of tapes. And they had an even bigger mouth in Gravano.

The Bull followed Michael Franzese, former captain of the rival Colombo crime family, into the dock to do for Gotti. Gotti got life. Gravano got five years and temporarily disappeared into a witness protection programme. (On 10 June 2002, Gotti, sixty-one, died of throat and mouth cancer at a prison hospital in

Springfield, Missouri.) When Gravano's secret cover was blown, he was exposed again to the danger of a hit – and the cops wanted one more squeal from the Bull. They wanted to clean up a bit more of Jersey.

'Al Certo is a family associate,' Gravano told the senators that April Fool's Day, a year after Gotti was put away. '[He] is with JoJo Corozzo, who is a "made" member [the favoured category of ambitious criminals allowed through their Italian genes to take an oath of allegiance to the Mafia] of the Gambino crime family.' Corozzo was going to take over from Gotti until he, too, went down.

Nobody was sure if Sammy was talking through his big backside. Maybe he'd done a deal with the same cops who'd spent years hunting him down and were now protecting him from his former 'associates' in the Mob.

Gravano named names like he was reading from the pages of the *Ring* magazine: middleweights Marvin Hagler and Vito Antuofermo, Francisco Damiani and Renaldo Snipes, heavyweights on the fringe of the big time. They all got a namecheck. There was no proof though, no concrete facts, no documents to link anyone to anyone else, just Gravano repeating he knew these fighters and their managers, as well as the top table of the Gambino family.

He sounded vaguely convincing. A fixed fight 'really doesn't happen any more', Sammy said, but he nonetheless pointed the finger of association all over the place. 'Our family [the Gambinos] has an interest in Buddy McGirt,' he repeated. 'Al Certo is a Gambino associate.'

It was a scattergun contribution that played well in the media. While it was entertaining enough, it lacked substance. But Sammy was the only card the cops had. Gravano then gave the hearing a short history lesson. Organised crime, he said, had 'walked away' from boxing in the mid-fifties, after Rocky Marciano retired undefeated as heavyweight champion. Was Rocky involved with the Mob, then? His manager of record was, after all, Al Weill, Carbo's 'associate'. The Bull wasn't saying. He'd said enough, maybe. Some monuments are better left untouched and dissing the Rock was probably more life-threatening that grassing up John Gotti.

But the Bull did allege that the Mob had been stirred into activity again, after nearly forty years out of the fight game; that made the panel sit up.

'Now,' he testified, 'because the size of the purses has gotten so big over the past twenty years, organised crime is more and more interested in getting back into it.'

This was a slice of baloney. The Mob never left boxing. They just didn't stumble about on the front pages any more, like Gravano and Gotti did. They played a longer, quieter game. They spread their interests. They cultivated friends in all quarters. It was naive to imagine the guys who'd grown up with boxing in their blood would walk away from the business just because Rocky, their biggest hero, an Italian American saint, was no longer about. Only a senator would believe it.

There are limpets still stuck to the body of boxing you'd not want around your place on a Sunday afternoon. There are pimps and drugs salesmen, money launderers and chancers, ex-cons and retired gunsels, former Mafia drivers, messengers, gamblers, gophers, touts and embezzlers, old numbers racketeers, hired muscle, creeps of all kinds. I've met them. Gravano, who will have met many more, was talking serious horseshit trying to persuade the senators the Mob had quit boxing. They might not move through the night like Hollywood hoodlums any more. They might not have names like 'Killer', 'Mad Dog', 'Fat John', 'Jimmy the Wop', 'The Weasel', 'Blinky' or 'Honest Bill'. But bad guys are bad guys. Crime is crime. That it was more obviously organised in the fifties did not mean those 'associates' retired to some rest home for worn-out gangsters.

Certo's reply to Gravano's allegations, without a nod to the niceties of judicial protocol, went to the point like a stiff jab.

'You guys,' he yelled at the Senate panel, in the unmistakable cadence of wop Jerseyese, 'pick out people with Italian names. You guys look at me with my dark glasses, and you see that I talk out of the side of my mouth, and you think I must be with the Mafia. I'd like to participate in a lie-detector test, and I'd like Gravano to participate in a lie-detector test, in front of all you guys, and then we'll see who's lying. I've never seen this man in my life. My dear daughter is dead, and I swear on her grave, this guy don't know nothing. He's full of shit when he says I know him.'

No more questions.

McGirt had earned a $1 million purse when he lost his title to the talented and temperamentally fragile Pernell Whittaker a month earlier, but had never otherwise been in the money. Certo's point was there was no sense in the Mob's tracking McGirt. 'Why would the Mafia nickel-and-dime us?' Certo asked, miffed at what he saw as the lack of logic in Gravano's allegations.

McGirt and Certo admitted Weiner had introduced them to Corozzo in Florida in 1991, but as unlikely as it might seem, they said they had no idea who he was. 'Stuart said he was an old childhood buddy of his,' said Certo. 'That was it.'

Now Al was putting puzzled looks on the senators' grave faces. They didn't know who to believe.

Weiner, obviously, was the target. He'd managed McGirt for twelve years, taking over the fighter's contract from his brother-in-law. 'That's how I got to know him,' Buddy said. 'Then he hooked up with Al.'

There was a good reason for that. Weiner did not have a manager's licence in any state in the union. Al did. They split 33 per cent of McGirt's money, a typical arrangement. That's how boxing has always worked. And nobody knew it better than Certo. Al was no innocent in the Garden. He went way back. He'd seen deals done nearly every day of his adult life. This wasn't the movies. It was business.

I could listen to Certo for hours. In fact, I did listen to him for hours. One afternoon at his tailor's shop in Secaucus, just over the bridge from Manhattan on the way to Jersey, we trawled through his past. He called out to someone in another room, 'Let's get some pizza in here for the Limeys! You like pizza?' If the senators couldn't get to the bottom of Gravano's claims, maybe I'd get a bigger picture from Al. And a slice or two of thin-crust pepperoni, no cheese thanks.

Not many people could provide such a good link between the old days and now, I thought. Most of them are dead, or took too many punches. Al shipped his share, but he lost just one bout in his whole career, amateur and pro. And he's hanging in there. I was fascinated to see how his reminiscences would stack up against the facts as I understood them.

Certo has a face borrowed from Burgess Meredith. You know he's 100 per cent 'boxing'. After he fought with distinction around New York in the early fifties, he earned his rep as a consummate cornerman and a good trainer and a manager. He's patched up more cuts than Ben Casey. Al describes the early years as incomparable to any that followed.

'Well,' he mused between chomps, 'in the late forties and the fifties, right after the war, that's when boxing flourished a lot, ya know. Fighters came home and they would fight in the Garden every Friday night. It was a joy. You would have the guys with the suits and ties and they would be with their wives or their girls . . . or whatever. It was a big night in New York. A big night. Great fighters.'

Certo saw the fight game from the bottom up.

'Well, ya know, I paid my dues. I had good people.There were so many good cornermen. And they were very helpful. They passed on the knowledge. I had a trainer, he wouldn't give you the sweat off his balls. I used to say, "Tooch, how do you handle a cut?" He'd say, "Take it easy, young man, take it easy, you've gotta eat a lotta spaghetti before you learn that stuff." He was a character. Ya learn, ya pick it up. Ya know?

'My record at that time was a good record; today it would be a super record. I'd be fighting for a title. Back then I was still fighting fours and sixes. It was nothing. I had around twenty-six fights and I lost one. [However many fights he did have, Certo would likely have been a match for modern lightweights.]

'If you were fighting in New Jersey in a main event like in Laurel Gardens, where I did a lot of my fighting, or anywhere else in Jersey, if you were an eight- or ten-round fighter and you wanted to get into the Garden and you wasn't such a big star or whatever-you would fight a six-rounder in the Garden . . . You would never get a main event.'

Laurel Gardens, the other Garden, was at 457 Springfield Avenue, near the intersection of 18th Avenue and South Tenth in Newark. It was a long way from Manhattan. Al never fought in the big Garden as a pro. But he knew its every contour, the back rooms and the corridors, he knew it like the chalk marks on the suits he made for a living. And he knew the men who ran it, although he struggles now to put names to some of the most infamous characters in boxing's history. Memories get punched about a bit in the business.

'The so-called wise guys ran boxing, gangsters or whatever you want to call them,' he recalls with matter-of-fact indifference, adding what was to become a repetitive theme of discussions I had with some of these old gym rats: 'But they were decent guys, they were men of their word. In other words they'd tell you, "This time it's not your night, kid, but next time don't worry about it, it's yours." Yeah, ya know, shit like that. Fixing didn't go on too much in boxing. It was . . . more or less . . . not a mismatch – but you were pretty close to knowing who was going to win.'

Al might have been reciting Marlon Brando's Terry Molloy speech from *On the Waterfront*, as the failed pug sat in the back of a cab with his mobbed-up brother and mulled over what might have been.

Cain and Abel had a better relationship.

Not so much fixes, then, Al, but fighters were imaginatively matched?

'Yeah. Very. Like there was a fight between Ernie "The Rock" Durando and Rocky Castellani. Castellani had Mob connections. Not Palermo or Carbo but they were on their level. I met Palermo and Jim Norris, I met 'em, yeah.'

This was of a pattern. Of the many people I spoke to in the course of writing this book, hardly anyone had ever come into direct physical contact with Carbo. No wonder they called him Mr Gray.

'Back in those years everybody was in bed with each other, but boxing ran smoothly and great and professionally and everything else. Then they had a big investigation into the IBC, and it kinda knocked them out of the box, ya know? I'm getting away from myself.

'Anyway, Castellani fought Ernie Durando at the Garden, in 1952, I think, and this was their second or third fight [Castellani beat the Rock on points in the Garden in 1950, and again in 1954]. It was on national television, the *Friday Night Fight*. Ernie Durando, from Bayonne, was a tremendous fighter. He was a dear friend of mine. Anyway, he hit Castellani with an uppercut and the photographer took a shot and his feet were off the ground. What an uppercut! Down he went. Boom! Ya know? It was a lucky shot. It was in the back of the *Daily News*. You could see Castellani flying. Oh boy. Anyhow, they stop the fight and Castellani's manager jumps in the ring and starts throwing a million fuckin' punches. He was hittin' every-fuckin'-body, including the referee. He was a boss of the Mob in New York. The story was that he'd bet $80,000 on the fight. So it wasn't a fix. Ya know? Ernie got lucky.'

That was an infamous fight. Castellani's manager who went beserk was a punk gangster called Tommy Eboli, also known as Tommy Ryan. When he launched into the referee, Ray Miller, not a lot of old ringsiders were surprised. And it didn't finish there. Later, Al Weill, notionally the Garden's matchmaker, even though Teddy Brenner had pretty much taken over from him, came to Castellani's dressing room, not the smartest of moves. Weill – French-born, New York-raised – had managed a string of fine fighters – Lou Ambers, Joey Archibald, Marty Servo – and was thick with Jacobs, Carbo and Norris. He might have reckoned he was fireproof. Eboli was enraged when Weill walked in and jumped on his back, knocking his glasses to the floor. Weill begged for mercy as Eboli thrashed into him.

The following day, Eboli had calmed down. But he knew he'd committed a grave error – not because it wasn't what he wanted to do, but because it might jeopardise future work in the Garden. So he fronted up to the Garden's main man at the time, Harry Markson, full of insincere contrition.

Markson had plenty of bottom, though. He looked at the quasi-repentant hood trying to ingratiate himself and said, 'You have sixty seconds to leave before I call the police.' Eboli left. That was the end of his career as a manager. Some time later, after crossing someone over a girl, cops found his dead body in a car boot, the classic Mafia 'off'.

'They was great years, ya know?' Certo continued. 'And to fight in the old Garden . . . even a four-rounder, even a six . . . wow.'

I was getting a take on how the fight fraternity could look in two directions at the same time. All that mattered, often, was the theatre of the occasion. What went on elsewhere, well, that was just part of the business. A guy lost a fight; Eboli finished his days in the boot of a car.

'The place was always packed. Always. Almost every seat was taken. The bouts were great. The Golden Gloves you couldn't get a seat at all.'

So, when did it begin to fade away a little bit? Was it later in the fifties?

'It was later, actually. Boxers always went down and came up but it seems like boxing today is way down, cos there's no trainers. As much as you like the televised bouts. In those years when we were fighting there was only eight or nine champions. To get a title shot? Forgeddaboudit! You had to know the President of the United States.

'So waddya do with all the champions today? Waddya say: "Hey, we don't want these guys," or what? "We got to cut back that work you've given us and go back to the eight titles and make these fighter lose all those pay days?" They're not gonna do that. Ya know? The boxing public. When I had Buddy McGirt all that time, HBO was riding high. If you were on HBO and you lost, that was it.

'I sat in the office of HBO with Seth Abraham and I said, "Some day you're gonna lose a lot of boxing and you're gonna lose a lot of shows." Ya know, to make the promo on these shows it costs quarter of a million dollars, maybe more. I says, "You guys have got enough money in the till to make two fights because if you lose the main event for whatever reason you can always use the other one and the thing keeps going." Because there was times when they lost a lot of money through postponing the fights cos somebody pulled out.'

Al would love to go back to the old days, mad mobsters and all. And the fans.

'Oh yeah, they were great fuckin' fans. They knew boxing. If you would look up into the balconies in certain clubs, like St Nick's, there'd be guys before the fight started playing cards, always playing cards, sometimes during the fight . . . Unless the fight was really good. Then they would stop and watch it.

'As great as television was for boxing, TV destroyed it, TV

destroyed boxing. They burnt up the talent so much they wanted any kind of title before they booked the fight. Ya know? ABC, CBS, NBC – if you were a fighter and you had a title, you were fine. They didn't care what kind of fighter you was. It didn't mean nothing. Ya know. So that killed boxing.

'In the fifties? It was flourishing. It was unbelievable. Unbelievable. Big fighters. In any era of boxing going all the way back to John L. Sullivan, the fifties I would say, up to the late sixties, was the greatest time in boxing there ever was. The greatest fighters that ever came out. Ray Robinson was the greatest fighter. There was two Ray Robinsons, maybe there was three. There was two Willie Peps, there was Sandy Saddler.'

Certainly Sandy might have thought there were two Willie Peps. In their four encounters, they laid bare their boxing souls, along with elbows, thumbs and rabbit punches, snarling and spitting into boxing history. Certo saw them all: '48, when Sandy stopped Willie in four; in '49, the *Ring*'s fight of the year, which Pep won on points; the rematch in 1950, when Saddler made Pep retire in the eighth; and the last of the rubber, in 1951, widely regarded as one of the dirtiest title fights of all time, Saddler winning in the ninth of fifteen. How could you not long for all that?

Saddler didn't pick up on the romance of it. In retirement, Pep, connected, white and darling of the Mob, was always called up to take a bow at big fights. Saddler, black, awkward and never quite fashionable, but who won three of their four fights, never did. You can work out why for yourself. For all its good intentions, boxing still is laced with the residue of prejudice.

'There was so many good fighters in the top twenty,' Certo continued. 'In the top ten any one of those fighters could have been champ. Any in the top twenty would have been the champ. I've been in the business for sixty-one years and I don't know who the champs are now. Boxing is a farce today. In every way. Right down to the corners.

'See, in my time, the fifties, there was no such thing as a cut man. You worked in the corner, you knew everything. You was a cornerman, that was it. Today they have a glove man, they got a bag man, they got a towel man, they're jackin' the people off. It's ridiculous, it's fuckin' ridiculous.

'Years ago they used to use stuff, New Skin it was called, and you'd put it on a cut, ya know, and you'd get like glass. It held the cut together but after the fight you had to pick it out. You'd lay down on a couch or whatever they had for you and you'd pick it out with the tweezers. You'd pick out the fuckin' glass. And there was a lot of other stuff they'd use, their own concoctions. You can't do that today. There's no magic. Don't let anybody bullshit you, there is no magic to it.'

Again, Certo makes it clear that, for all the malarkey, the old days were the good days – even if they were also the bad days. No senator was going to understand that.

'Let his soul rest in peace, Al Gavin, he became a real big cut man because he worked with the heavyweight champion, but he didn't know his ass from his elbow. I knew the guy a long time.'

Gavin was special, though. He had a knack. Magic, if you like. There were few better corner medics in boxing. He was born in Brooklyn to boxing-mad parents from County Wicklow. Gavin was also a realist. He had twenty amateur fights, winning fourteen, and discovered he had no punch and a so-so chin. When he went to the influential Al Braverman, he was told to forget it. This was the fifties. Ray Robinson was on the loose in Gavin's weight division. 'I looked at some of the other guys, like Gene Fullmer and Rocky Castellani. No way I could beat fighters like that,' he told Thomas Hauser. So Gavin became one of boxing's great cornermen. Among his clients were Lennox Lewis and Naseem Hamed, as well as the greatest bleeder of them all, Chuck Wepner. 'It's a combination of art, science and luck,' Gavin said not long before he died in 2004. 'I'm nothing special. I just go out and do my job. I'm not a big shot.

I'm just a guy who likes boxing.' According to most, though, Gavin was a magician, the Swami of the Swab . . . But Certo was having none of it.

'There is no magic. Pressure, pressure, pressure. That's what it is. They give you stuff called adrenalin 1001 [that's a thousand parts water, to one part adrenalin]. It's not enough to do shit. All it does is keep the cut clean. That's the only reason I use it, more or less if the cut is clean I'll just apply pressure on it.'

Did they ever think of stitching them straight away in the fifties?

'In the old Garden and St Nick's if you got bad cuts the doctor and the commission doctor would stitch you right there in the dressing room and there was no needle or no nuttin'. It was, "Just stitch me up!"'

Getting stitched up has probably changed in the literal sense but not metaphorically in the boxing business. Clean on the outside, still dirty inside.

Certo bemoans other passing traditions. He reckons – and he's probably right – there is not the attachment of the fighters and trainers to the writers any more. The bond was broken a long time ago. Now there is merely suspicion and paranoia on both sides.

'There were some great writers, ya know, years ago. They used to go to the gym. Like Teddy Brenner, when he first came into the business, he was a young guy and he was a great matchmaker, right? No. He wasn't a great matchmaker: you could pull two guys out of a hat and the fight was a great, great fight. Teddy used to use the opinions of the writers and they would give him their advice. The writers made a lot of the matches for the old Garden, ya know? They'd say this fighter's good and that fighter's good. The matchmakers would pick their minds, ya know, they were the experts. They lived in the gym, the writers. Today you never see them in the gym. They don't even know what the gym fuckin' looks like.'

When they weren't at Stillman's or Gleason's, or hanging

around the Garden, the writers and other fight junkies were at Toots Shor's, the bar where all the world was distilled into a few anecdotes. Bernard 'Toots' Shor used to sell underpants door to door before he came up from Philadelphia in 1930, and got work as a bouncer at a place called the Five O'Clock Club. That would be the first five o'clock, when patrons wandered home. Toots made friends with a load of celebrities and opened his own bar on West 51st Street, right near the Garden. It was a natural.

'Oh, that's where everything happened, a lot of great fights were made there. He was a character, ya know. Him and Sherman Billingsley, who owned the Stork Club. They would go to each other's restaurants but Toots Shor's was more known for the sports figures, ya know. Boxers, baseball players used to hang out there. Rocky Graziano, who was my dear friend, may his soul rest in peace, he had a fight with Jackie Gleason. He was drunk and Rocky was feeling good too, but Rocky wasn't a violent guy, ya know, but Jackie was breaking his balls at Toots Shor's. "You're a punch-drunk fighter," and all that bullshit, ya know. "You guinea bastard." Rocky went whack! Knocked him cold. Right in Toots Shor's . . . Jackie never had Rocky on his show after that.'

Certo, meanwhile, had to earn a living. He had his tailor's shop in Secaucus, but he missed the fight game.

'When I stopped boxing, I moved from Hoboken to Secaucus and I liked working with the kids. I worked with the PAL, ya know, the Police Athletic League, and I got a programme started in Secaucus and we got our ass kicked for a few years but then I came out with a good team. I also opened up a gym, a local gym, it was an old theatre house and I rented it out. I wanted to make a break in this growing business out here, ya know, and there was times I was doing real good. Real good.

'In fact, this place was known as a hangout for ex-fighters. We used to have shows upstairs. Anyhow, all the champions used to come here. I used to make clothes for them. Being an ex-fighter, they knew me, I knew them. Rocky Graziano was close to me, Marciano was very close. In fact, he got killed in one of my jackets. Jersey Joe Walcott was my close friend. They were my idols. Great fighters.

'Upstairs was my showroom and Sonny Liston come in and he had all the reporters following him and they all knew Al Certo from Secaucus was a boxing guy and a former fighter and all that bullshit, so they stopped in and he was looking at all these pictures and he said, "Where's all the black celebrities?" Well, I had a few up there so I'm showing him and then one of the reporters says to me, "We've been following him from Journal Square [in Union City, New Jersey] all the way to here, that's how far he's walked, about three or four miles, and we can't get a word out of his mouth!" So I says, "Sonny, why don't you talk to the guys? They're trying to do a job, come on."

'So I call the guys over and I says, "Ask him what you want." And a guy asks Sonny, "Well, what do you think you are going to do with Chuck Wepner?" And he was looking at the pictures and he said, "I'm gonna bust his fuckin' head." Just like that, and he said nuttin' more – and that's just what the hell he done.'

What did you reckon to Sonny?

'He was a killer. Probably his hands were tied in the fight with Muhammad Ali, ya know.'

The second one?

'Yeah.'

First one on the level, then?

'No. The first one, Muhammad Ali didn't even want to come out of the corner for that fight. He almost blew it. But for some reason, ya know, Sonny . . . He said he had something in his eyes, which was bullshit. [It was Ali who complained he had a stinging grease in his eye after rubbing up against Liston. Sonny pulled out with a ricked shoulder.] And ya know they stopped the fight.

Maybe the Muslims got to him. "This is one fight you can't win." . . . Ya know?'

And the second fight?

'The second fight was a tank job.'

José Torres reckons it was on the level.

'José got hit with a lot of punches, ya know. He's a dear friend but he got hit with a lot of punches.'

We reminisce at large, about José, about Floyd Patterson, who shared a gym at Cus D'Amato's place.

'Patterson. There's my dear friend. Ya know you've got a cousin or something and you know he's good, right? Patterson was close to me, ya know? The last time I seen him he didn't know who the fuck I was. We were at some kinda function. This was about three years ago.' [Patterson had Alzheimer's at the time. He has died since.]

'When he was the commissioner I sort of helped him, ya know. Ya see a lot of fighters, they don't know too many things about boxing, so they need help. Jersey Joe Walcott was another one who was a super guy, and him and I was very close, but he didn't know his ass from his elbow. There's a lot to boxing. So I had to help Floyd when he was the commissioner and we would go to different places like Connecticut and Boston. He didn't last too long. Even when he was the commissioner he was losing it.'

We talk about Buddy McGirt.

'Buddy was an exceptionally good fighter. Who was the British fighter Buddy fought and he made a mess of him? Gary Jacobs. Right. That poor kid. Gary was a good fighter but Buddy was just too hot for him. He punched the shit out of him. That was in the Felt Forum in the Garden. Two weeks later we went in against this giant from Argentina. Buddy was something. It's a shame when we fought Whittaker. That fuckin' name – it aggravates me. Cos I know what Buddy would have done to him. He would have knocked him out. If we would have had two hands we would have knocked him out.

Guaranteed. He had no left hand and his right hand was shakin', and he was already fuckin' his brains out with the broads, ya know. Two years prior Buddy was slippin' . . . but that's another story.'

What about in the old days. Did the fighters get among the women?

'Well, boxing was too tough then. You couldn't fuck around with broads, you'd get yer head hammered in. Every fight in the Garden it was a murder. You had to be up for it. So you couldn't afford to be a playboy, ya know. Today they think of nothing of it. Buddy used to come here after a fight and he would have to meet a broad at twelve o'clock and I'd say, "What are you doing here?" Cos he lived in Long Island, ya know. And he would come back at three o'clock and meet another broad and then go home. I used to say, "You're fuckin' your brains out. Are you nuts or what? Let's make some money first."

'He fought about two or three years beyond his time, ya know, but he was a great fighter. He learned everything from me and from the guys hanging around, Walcott, Willie Pep, Giardello – they all loved him, he was a great talent. And I used to tell him, "You listen to these guys when they talk. You get great knowledge. And use your own knowledge."'

I wanted to know more about the fifties from Al.

'The fifties was the greatest. Don't let anybody try and tell you different. But there's nobody left. Willie Pep is in a nursing home. [He died in a nursing home in Connecticut in November 2006. Dementia pugilistica had long since broken his mind.] The only dear friend I have left is Joey Giardello. He was a great, great fighter. He comes around every once in a while. Although he don't drive no more . . . [Giardello died in September 2008.]

'I done a lot of work for a lot of stars. Sinatra was the biggest guy I worked for. He was from Hoboken. He was my idol and his family was in the fight game, too.'

So, tell me about Frank and his father. He fought as Marty O'Brien, didn't he, a bantamweight?

'Yeah. See, in them days, back in the twenties and thirties, you had to be Irish or Jewish. Champ Sieger was Marty's brother-in-law. He was Dolly Sinatra's brother. Dolly was Frank's mother. Right? He fought under the name Champ Sieger. Sieger was a Jewish name and Marty O'Brien was Irish. Ya know? And ya know what else? They fought each other.'

The records show Dolly's brother boxed as Young Sieger, but maybe he had more than one name. He'd hardly have been alone. It was a long time ago. Sieger knocked Frank's dad out in seven rounds the first time – it was one of five wins in twenty professional bouts – and was disqualified in the rematch.

Marty O'Brien – born Anthony Martin Sinatra in Catania, Sicily, in 1894 – was not going to make it as a world champion, but his son had a voice that would make him a legend in a different part of the entertainment forest. Frank never cut his ties to his Hoboken and Italian roots, but, according to Al, would not have made a fighter even as ordinary as his dad. Sinatra Sr won only one of his six fights stretched over ten years.

Sinatra Jr was mobbed up in the wider sense of the term, because, through choice or the fall of the cards, he had to accommodate their presence when he became famous. In fact, there wasn't a choice.

Was Frank any good with his fists? I wondered.

'Nah! Nah! He didn't know his ass from his elbow. He wanted to be a tough guy but he wasn't up to it. He wasn't tough. In fact, one of my cornermen who worked with me a few times, he was an old-timer who worked with all the champions, Al Silvani, and Al Silvani was Sinatra's valet, his bodyguard. Ya know? He knew him real well.'

Silvani was a fascinating bit player. He trained several world champions – Marciano, La Motta, Basilio, Graziano, Henry Armstrong, Fritzie Zivic, Ingemar Johansson and Lou Ambers among them. He went on to be an actor, appearing in *From Here to Eternity*, *Ocean's Eleven*, *Robin and the Seven Hoods*. The fact he drove for Sinatra probably didn't hurt. Silvani died in 1996.

'Al said Frank wasn't tough.' One Al said it to the other Al; he almost certainly didn't say it to Frank.

But Frank could sing. They all agreed on that.

And, just as Frank got out of Hoboken and reached for legitimacy (never convincing respectable society of his bona fides), so Certo once harboured thoughts of a life not necessarily allied to beating the crap out of the Irish, the wops or the Yids.

'Living in Hoboken, coming from a tough neighbourhood, I used to fight all the Irish, ya know? They used to say, "You guinea bastard!" Ya know? And I'd say, "Fuck you! You Irish bastard!" I was ten years old. The cops used to call me that: "Get the hell out of here, ya guinea bastard!"

'I've always been a tailor. My grandfather was a tailor, and my father. It's in the family. I kinda never wanted to be a fighter . . . I wanted to be a tap dancer. I loved dancing. My father was in show business, he had his own band, ya know? In Jersey.

'There was a place in Journal Square where you could learn how to dance, but you had to enrol and be a fuckin' ballet intern. I didn't know that. Who the hell wants to be a fuckin' ballet dancer? If the guys in the neighbourhood would have learned I was a fuckin' ballet dancer I would have had all the fights I wanted. Oh my God!

'So we kept it quiet for a coupla weeks but then it came out and, sure enough, that was the end of my tap dancing. So I got married and took up boxing. I won the Gloves in the amateurs and my manager told me to turn pro and I said, "All right, I'll try it."'

Al then came up with the craziest idea.

'When I came out of the amateurs I wanted to use Frank's name as a stepping stone. I was twenty-two years old. I thought I was old already, ya know? I gotta get a good firm foot in life. I was working at a family car wash over in Hoboken. I was cleaning

Sinatra's old man's car. He was a fire captain over in Hoboken. "Marty," I said, "Waddya think? Do you think he'd be mad if I used his name?" "Waddya mean?" "Well, I just won the Golden Gloves and I'm looking for a push, ya know? I wanna use Frank's name." "Did I hear you good? What the fuck you wanna use somebody else's name for? Look at me, Al. Look at me. Marty O'Brien. Nobody even knows me."

'So . . . that didn't happen. When my career was over – ya know I got hurt, I cut my hand – although that's another story . . .

'Jake La Motta was another dear friend. Jake had more fuckin broads. Jake and Rocky [Marciano] and I was real close. Jake was getting married, I think for the sixth time, I said to Rocky, "Are you going to the wedding?" Rocky said, "No. He can't make up his mind. Forgeddaboutit." Rocky – let his soul rest in peace – was a character, so fuckin' cheap.'

Is that the way he grew up?

'I guess so. They were hard times.'

How good do you think he was?

'He wasn't my cup of tea as a fighter but how could you dispute what he did? He had a cute way of boxing – when a punch came out to him he would pull all the way back so when he got hit there was no power in it. The times when he did get hit – and Walcott he got hit flush . . . But he wasn't much of a talker.'

Ali would have beaten Marciano, wouldn't he?

'I think so, but I did say to Rocky, "How did you think you would have done with him?" Rocky was a type of guy, he was a funny guy, he always said the thing you wanted to hear. He never told you the truth. So he says, "Al, I think I could take him now."

'You're a young boy.'

Thanks.

'Do you remember when they used to have the black-tie affairs over there in England? Who was the main guy?'

Jack Solomons.

'Solomons. Right. Do you remember the great fights they used to have? Why is it that over there in England a fighter is like a fuckin' golfer over here? You know they are big stars. In America they look at fighters like hoodlums. Like shit. I think it's got to do with where boxing came from. It's changing a lot, it's not like it used to be. We don't get the amateurs any more. There's no trainers around any more.'

Al slipped from smiling and being high on the past to melancholy. The past all of a sudden seemed a long, long time ago.

'Television killed boxing. Do you know there was a period in the fifties when boxing was on televison seven days a week. National television, seven days a week. *Gillette. Friday Night Fights*. I can't remember all of 'em.

'At that time there was a million fighters, sure. Like in the fifties I fought in the Gloves and at that time in my division there was 2,000 fighters, 135lb or whatever, and that was just in my division, and at that time for me to win the title I fought eight times in a row without taking a bye, three twos. When you fought in the final it was three threes.'

Did everyone turn pro in those days?

'I would say on an average, let's say, out of five hundred kids, fifty good kids would come out of it and turn pro. You don't get the characters any more either. You remember Al "Bummy" Davis? Got shot outside of a bar . . .'

The names weren't flowing so easily now. We'd been pretty much through the card. So I asked him, how good was Billy Graham, one of the renowned fighters of the fifties who never won a world title?

'Billy was my dear friend. He used to come here all the time. We used to love to hang out together and talk about old times. If you lost to somebody you'd say, "You should have done this or that." Ya know, break balls. But Billy, if you look at his record, he told me he had forty-three four-rounders before he fought six-rounders and eight-rounders. Imagine that.' [It was sixteen. Graham started out in 1941 and within a year he was matched over six rounds against Harry Diduck at the Broadway Arena in Brooklyn. Billy was incredibly active down the bill, though, and fought his first eight-rounder in his thirty-fifth fight, at St Nick's, in September 1942, just twenty-one months after his debut.]

He should have been a champion, shouldn't he?

'Well, he should have beaten Kid Gavilan but they fucked him out of it.'

That's one way of putting it. Was Graham bitter about that at all?

'He was kinda bitter. Who the hell knows? At that time the smart money was on Gavilan. The fix was in. But who the hell knows?'

Billy never said?

'No, he never said.'

Don't suppose he'd know.

'No, he wouldn't know who's doing what . . . Who'd know? Ya know? . . . Better take my pills. Otherwise I won't be able to get a hard-on tonight. Heh! That's what memories are for, right?'

17

LOU DUVA, A FIGHTER

Lou Duva, who has trained many world champions, has a fighter's recall of the old place, the Garden. I caught up with him at his gym in West Paterson, New Jersey. He doesn't give up smiles easily, but Lou, for once, is in expansive mood. Maybe one of the loudest men in boxing is getting soft. Or maybe not.

'First time at the Garden? It had to be the late thirties. My brother was going to box a four-rounder. At that time, you had to have about forty or fifty fights to even fight a four-rounder in the Garden. So I went over there with him. Unfortunately he didn't fight. But they had a system where, if you didn't fight, they'd put you on the next show. I think they gave him something like $75 to be on standby. When he finally fought his four-rounder, I think he got $125. And I think he won the fight. [Carl Duva fought in the Garden on 8 March 1940, and lost on points over four rounds to Eddie Carroll. It was Carl's last fight.]

'That was good money in them days, oh yeah. People were taking home maybe $30 a week then. I was born in New York, in Little Italy, down Mulberry Street, the real immigrant section. I came over to Jersey when I was four years old. I got involved in boxing when I was ten years old. That came about because of my brother. My brother was working in the dye house. In those days, you didn't have any luxuries. Today, young guys don't want to fight.

They got investments and that. If a guy's got any smarts at all, he don't wanna fight. He has to have a new car, he has to have an apartment. He has to have all those things, you understand, that guys in the old days never did.

'Most of the fighters I grew up with, they all had jobs. Very few didn't have a job. Let me put it this way, guys that were owned – or handled – by gangsters or big restaurateurs, stuff like that, could afford to keep goin', and they'd work out in Gleason's gym, or Stillman's gym, over on Eighth Avenue. Ordinarily, a boxer at that time could box somewhere just about every week. So they'd get in the ring for $150, $125, some times a couple of hundred dollars for a main event. But they'd work during the day, in a factory, drivin' a truck, or do somethin'. And they nearly all trained at night, some trained at a regular gym, some trained in the back of bars. Sometimes the regular patrons drinkin' there would watch the guys fight – Manhattan, Jersey, all over.

'So my brother'd come home about seven o'clock, I'd pack his bag and go to the gym with him, down Paterson on West Broadway. I'd watch other fighters and they'd show me stuff, how to use a jab, messin' around. So that's how I got involved in boxin'. Jersey's a home of great fighters. Great fighters . . .'

Did you see the big names around at the time?

'Sure. Damon Runyon ran a couple of fights. Sugar Ray Robinson boxed Charlie Fusari at the Roosevelt Stadium in Jersey City [in 1950].'

That was for Ray's world welterweight title, a fight in an unfashionable arena, away from the Mob. Fusari was a good fighter, going into the bout with a 63–7–1 record, but it was almost a sideshow. Robinson won easily on points over the fifteen rounds. It had more credibility than his next three fights, however, a one-round knockout against Jose Basora (who went down four times) for an obscure version of the world middleweight title, and two non-title, marking-time money fights, easily won, against Billy

Brown and Joe Rindone. Robinson had the luxury of being great, fighting who he wanted to fight, when and where he liked most of the time, and telling the Mob to go to hell.

Back in Jersey, they were just pleased to see him that night, and Lou was there to cheer Charlie on, which didn't help much, as he struggled to the bell.

Where Fusari was from, though, there were plenty of others, queuing up for their chance. 'There used to be a lot of good tough Jersey guys,' Duva said. 'What we used to do, a lot of us, is we used to go up behind bars – they called them smokers – and you would get ten or twelve dollars if you won and eight dollars if you lost, or they'd give you a watch and you'd get the money off that, and you could box amateur and then you'd box under a different name. I used to make myself maybe thirty dollars a week and I'd go and box two or three times, exhibitions and stuff like that.

'We used to have a guy by the name of Jack Lock as a manager and he used to pile eight or ten guys into a car. We were supposed to get three dollars a day for food. We never got food – he'd make sandwiches for us. We always used to have a standing joke after the fight driving back home. "There's a good diner down the road," he'd say. We never found that diner and consequently he kept the three dollars.'

But the guys wanted to fight regardless, didn't they?

'Always wanted to fight. It was, "Where's your fight tonight? Let's go down to Trenton. There's one in Newark, in Brooklyn." There was always a boxing show going on. It's not like today. Everything's controlled with the television.'

And the Garden was the magic place to go.

'Absolutely. That was the Mecca of boxing. I used to go to Europe a lot with my boxers and whether it be in Italy or London they'd always ask about the Garden. Fighters used to say, "One day I'm going to box in the Garden." When you boxed in the Garden you got the recognition. You'd arrived as a fighter.'

You had to be good to box there.

'Absolutely. But you had to have connection too.

'Let me tell you where I think it was better in those days – and most of the old-timers agree with me. If a fighter was supposed to be training, the trainer was responsible for him. He trained him, he taught him. Nobody came over and told him what to do. No investor came over and said, "Wait a minute, you've got to get the right uniform or the right music," or this or that.

'It was a three-step programme. You had the manager, you had the trainer and you had the fighter. That trainer trained the fighter, that trainer gave him a massage, a rub-down. They don't do any of that today. I mean, right now, even with a good trainer, there's very few of them out there that's really close to a fighter, who do that stuff. Today they've got lawyers, accountants, musical directors. When you talk about an entourage, they've got one.'

Fighters didn't have an entourage in the old days?

'No, you was a fighter, period, that's it. A trainer, a manager and that's it.'

How many fighters would a trainer have normally?

'They would have maybe a dozen. He would have a couple of assistants working with him, to help with the gloves etc., but the main guy was the trainer. Not only the trainer but he was a teacher. Today they've got nutritionists, personal trainers, every goddamn thing under the sun. When you're overlapping personalities something's got to go wrong.'

Tell me about some of the fighters you saw fight in the Garden in the fifties, Lou.

'You had guys like Sandy Saddler, Tony Graziano, Tony Zale and Marcel Cerdan, you had Jake La Motta, I mean, you had fighters in those days and they fought each other and there was no personalities out there, you know?'

You knew Rocky Marciano pretty well, didn't you?

'My main guy. He was my personal friend but one of the

cheapest guys. Though he could be generous too. I've got pictures of Joe Louis and Rocky [Marciano] playing golf. I hired Joe and Rocky for 25 hundred dollars apiece for a week when I was promoting in Atlantic City, and they'd come down and play golf with the politicians for 25 hundred dollars apiece. Rocky set it up. For Joe.

'Barbara, his wife, was pretty tough, a tough Irishwoman. She kept Rocky in place, and half the time Rocky didn't want to go home.

'I knew Rocky and all the guys. This is how I got to know the boxing business. I was driving a truck at the time and I started six in the morning. I used to get in at the same time as the garbage men in New York, 34th Street and all that, and I used to drop off my load. I got in with all the elevator operators and I would get up early to bring my stuff. The reason I did that is because then I could go down to Stillman's gym later. I'd park my truck and make a connection with the guy that had the parking lot there, then walk over to the gym, and go upstairs. I got to know everybody over there. I started to watch. But I didn't watch the fighters. I watched the trainers and managers. They had a bank of telephones over there and calls would come in at the gym. It was, "I'm looking for a six-rounder. Hey, maybe I can help you." And you'd see them stealing the bouts. Then you watch the trainers.I'm talking about Ray Arcel, who was a master. Manny Seaman, Johnny Sullo [Joe Miceli's trainer], all those guys. And I watched them.

'One of the guys I loved was Charley Goldman. He used to take care of Rocky Marciano. He took a raw product and he taught him. He would tie his feet, he'd tie his hands. He'd make him carry a towel under his arm. He would go to punch, Charlie'd say, "Pick that up with a glove." He'd teach, and you don't see anybody doing any of that.

'We used to go up Joe Louis's training camp too, up in Pompton Lakes, New Jersey. Joe put that little town on the map. I'd caddy golf at the country club in the morning then after about eleven o'clock I'd go on to the highway and hitch a ride up to Pompton Lakes. He was some guy, wasn't he? Great guy. They got to know me up there. The manager, George Gainsford, was a good friend of mine. They'd give me a sandwich, I was just a kid. The house is still up there, in fact. Newspapermen were always there. Not like today. Today they don't cover nothin'. Everything is by computer or phone and once in a while they have a press conference.

'Back then, when the fighters trained up in camp, they always had the press there. In New York, at Stillman's, those guys were always hanging around the gym. In the afternoon when they were done they would go across to the Garden, the Garden cafeteria, they'd all hang out there and we'd go over and they'd exchange stories – "You should've seen that fight we had in Brooklyn." We exchanged tales with the newspaper guys. So that was boxing. You haven't got any of that stuff any more.

'I would go to see Joe Louis train, and I'd grab a newspaper guy up there, let's say Lester Bromberg or somebody. "Hey, listen, can I ride up with you up to Marciano's?" And they'd say, "Sure, come on, I'm going up there now." He trained in the afternoon. When he'd done with Rocky, I'd jump another ride, go up to see Sugar Ray Robinson up at his camp. I'd get home by nine, ten o'clock. But that was my life. I'd do that five times a week.'

And the boxing writers knew their business?

'Absolutely. I'll tell you one thing about the writers in those days; they were the best allies a promoter or fighter could have. They told in-depth stories. They had the reality, they built the fighter up, or they tore a fighter down, but they were always out there doing their job. If they had to I'm sure some of them would even make up a story, just to push a fight. I was with Jimmy Breslin when he got hired for fifty dollars a week by Rocky Marciano and the name of the show was "The Main Event".'

I'm not sure Lou meant to tell me that in the way he did.

Because it only confirmed the sad suspicion I had that some underpaid hacks were picking up chump change, rolling over for the promoters, pushing product rather than digging for real stories. But Lou can see it only from the other side of the fence. And he called his company Main Event.

Did you know Carbo?

'Yeah. I never had any problem with him. I know he was real close with Rocky, I know he backed Rocky up in his restaurant over there on Madison Avenue. Now? None of that. Fighters are all going to court now. If you can't be a fighter become a lawyer.'

What kind of percentages was Carbo taking? Ike Williams for instance – he ended up in big trouble financially.

'Let me tell you something about fighters ending up in big trouble financially. They lived from fight to fight. They did then and they do now. The thing that a fighter needs more than anything, once he can fight, once he knows he can fight, and once the people know he can fight, and he's got some career in front of him, he needs a good lawyer or somebody that's a family member that's close to him who will tell him what's the right thing to do. I'll show you football players around that have no money and they were making millions of dollars a year playing football. It's their lifestyle. From one game to another game, they don't control their money and they don't invest their money right. They think it's not going to stop. I had Evander Holyfield for fourteen years. He made $200 million with me. He's in trouble. He had about three or four wives and kids. He shouldn't still be fighting. That's why we walked away from him. I thought we had a deal that he was going to retire and I said that, but he went back to the ring because everybody started bullshitting him and he had too many guys working for him. I didn't bullshit him, I told him the truth. He went with King and King brought him along and he's still fighting.

'Fighters, they're the worst culprits, fighters themselves. They talk about guys like Bernard Hopkins who was in jail for five years [for 'strong-arm robbery'] and been through all that stuff and still I think he's right in what he does, he's going to demand money, this and that. He stood up for what he thought was right, he did his own thing. But there's some of these other guys they've got no clue at all as far as lifestyle is concerned.'

Are they too impressionable?

'Oh, they run away with themselves. They don't give anything back. I get my fighters and we go to schools and we talk. We bring them in here and the other gyms I used to have. I have cops, they get some kids and bring them in here. They're ready to go to jail, some of them, stealing cars, whatever. Bring them in and we teach them. I've got scholarships for guys that became lawyers, doctors, etc.

'I'm eighty-three [this was 2004], that's a long time in boxing. I'm getting near the end now because it pisses me off when I see some of these guys coming in here thinking they know about boxing and they don't know shit and the way the sport is changing. I mean, when you see television knocking boxing off the air. Pay per view is fine for a few boxers. But very few. No one knows who the fighters are any more. I'd like to see a national commissioner, I'd like to see boxing people handle it. I don't want to see politicians being appointed or judges or referees. I want to see fighters, people involved in boxing, former trainers and stuff. They've got no place to go and they don't give anything back. The evil of everything is retirement. The word retirement: I hate it. And if you don't do anything with your life you're going to die fast and die old. Give something back. Help the kids. Visit the hospitals. I'm in the ring, I'm out the ring and they're going to have to carry me the hell out.'

Are there any good guys left in boxing?

'There's a lot of good people. But the era that I liked, where you really had to show what you were all about was maybe the forties, when I was boxing amateur and I forged my birth certificate from seventeen to eighteen so I could get into the army. I was boxing and an instructor and I learned what life is all about. They were the tough times. And the best. That was the real time as far as I'm concerned. The forties and fifties.

'I love the business, I love boxing. What pisses me off is when it's not done right. I like to take a kid and develop him. I have kids out there now if they weren't over here in the gym they'd be serving time for murder. Two of them I send to school to learn. I've got a kid, took him from Brooklyn and brought him over here, made him a trainer, and he just graduated, became a paralegal. That's one notch below a lawyer. I've got a couple of other kids going to school and some satisfaction that I gave something back. And I don't know for the life of me why anybody's in the position to do it shouldn't do it. All athletes, I think the worst trait they've got is they don't give anything back.

'The worst time in a boxer's life is when he quits boxing. It's the worst time. Where does he go? He's playing golf. "I don't know what to do tomorrow. Maybe my wife will take me shopping." Come on. You don't know what to do with yourself? There's so much in life, give something back.

'Meldrick Taylor. He's working round the corner with the Muslim or church people. He was a world champion. He blew most of his money, millions of dollars. There's so many ways to get lost today. The worst sin you can commit is trying to con a guy instead of telling him the truth. Boxing is a good sport but it's a tough sport. If you want to see good there's a lot of good in this business. If you want to see bad there's a lot of bad in this sport.'

I left Lou's gym not altogether wiser, but feeling just a little better about things. The old curmudgeon has made enemies, mistakes, fallen out with family and friends – but he's still in there. He's a fighter.

18

A COUPLE OF ARTISTS

LeRoy Neiman came to New York as a romantic artist would, with his eyes wide and his mouth open. Both have stayed that way, more or less, ever since.

LeRoy was in awe of the place. 'Every day in New York was just a great day,' he says. 'I was in awe of the place.' And I'm sure he reckons New York was some times in awe of him. He does not shy away from letting his publicists describe him as 'probably the most popular living artist in the United States'. You can bet his scraggy-trousered mates back in St Paul, Minnesota, in the thirties and forties would have raised a chuckle over that.

'I would say St Paul was a bit like Warrington in England. Industrial, a very tough town,' LeRoy says, in surroundings that do not evoke images of the Depression. We are sitting in his quite sumptuous apartment in a rococo block on the west side of Central Park, and clearly not some cold-water tenement. The apartment stretches for some distance. The doors are thick. The ceilings are tall. The paintings, even those not belonging to the resident, look expensive. LeRoy has done well.

He speaks to you. Like this. In short, almost asthmatic bursts. Every sentence a sketch. It is quite dainty. His delivery is hypnotic, like Sugar Ray's jab. He peppers you with facts as he would a canvas with a fine brush, touching up the final product before sale.

After leaving St Paul and embracing Manhattan when it was still mostly in black and white, he went on to carve a unique niche in boxing. Essentially, Neiman became everybody's friend. He went to fights with Angelo Dundee. He entered, unchallenged and without the accoutrements of press or promoter, the Garden dressing rooms of famous fighters only moments before they were to do battle. He knew wise guys. He got good seats. He sat with celebrities. He was a celebrity. He was a chimera, a ghostly but undeniable presence.

LeRoy still wears a self-conscious, Daliesque moustache and has the air of an outrageous poseur, but he is easy to like, even if it is hard to imagine him convincing the fight crowd he was one of their own.

Stan Isaacs wrote in *Newsday* in 1968: 'Whether one approves of Neiman's work or not . . . one must agree that he is a work of art himself.'

The artist agrees.

Yet he did strike a chord among fight folk. Maybe it was because it is a milieu of eccentrics. It has always been where he feels comfortable, a coming together of two widely disparate art forms that appeals to his wild nature. Neiman, when he got to New York via London and Paris in the late fifties, was a fan and an astute observer. He knew boxing and he knew about tough guys. Still does. He is misleadingly effete.

He said of himself once (possibly many times, come to think of it): 'I guess I created LeRoy Neiman. Nobody else told me how to do it. I'm a believer in the theory that the artist is as important as his work.' LeRoy is the art world's version of Bert Sugar.

How did such a towering ego not get his block knocked off at some point in Toots Shor's, Lindy's or the foyer of the Forrest Hotel? Perhaps any prospective antagonist would feel ridiculous launching an assault on someone seemingly not on level terms with the fight world. And maybe the guy has a left jab up his sleeve.

'During the Depression in St Paul, they had boxing progammes in school, which were designed to keep kids off the streets. Mostly, violence in the streets involved fist fights, not with weapons like today. So we all regarded boxing as civilised. You had gloves. Discipline. Referees. St Paul was a boxing town. Like Philadelphia. We had one guy, Mike Gibbons, the big man in town – he was called the St Paul Phantom, because you couldn't hit him – he had a gym there. And there was his younger brother. Tommy. They were legends. All Irish.'

Tommy, unwittingly, was part of one of the great boxing scams of the twentieth century when he fought Jack Dempsey at Shelby, Montana, in 1923. The fight infamously bankrupted the small town and the combatants left with considerably more urgency than they'd shown in arriving. Dempsey's manager, Doc Kearns, and his promoter, Tex Rickard, would go on to attach themselves like limpets to Madison Square Garden and all it had to offer. Gibbons didn't beat Dempsey – very few did – but he gave a good account of himself and, at the end of a long career, had a record heavyweights of today can only dream about. He boxed through the weights, often taking on much bigger opponents. He lost just four out of 106 fights. He was knocked out only once. Tommy Gibbons was a special fighter.

Neiman, thankfully, loves to reminisce: 'We admired proper boxing, like the Gibbons boys showed. Which is why we always wondered as kids why our fighters would go to New York and get knocked out. To us, that was brawling. In cities like St Paul, you didn't always say what you meant; when I came to New York I discovered that wasn't the case here. You can get anything you want here. But you've got to ask for it.'

Before New York, Neiman tried London. He remembers seeing Terry Spinks there in the fifties. In New York, he palled up with the likes of Rocky Graziano. LeRoy was not coy about moving in interesting circles.

'Rocky was a genuinely tough Lower East Side kid. Like Tony Curtis, Jake La Motta, and some of those guys. If they weren't tough with their fists, they were tough in every other way. Hard guys. Often the referee was as tough a guy as the two fighters. So were the crowd. You were swimming against the sewer there, Rocky would say. Nobody trusts anybody from outside. It was a great, close-knit community.'

And pretty touchy about its image, sometimes.

LeRoy recalled an intriguing incident involving the management of Madison Square Garden. 'Go to Room 200, you'll see it's still there, one of my paintings. Take a close look at it. Before the fights, the mayor and all the big shots, the freeloaders, that's where they went before a big fight. Room 200. I did most of the paintings in there. The Garden at the time commissioned me to do a painting of fights at the Garden. It was a huge painting. The ring, everything. Jack Dempsey was alive, Joe Louis was alive, all the champions, Johnny Addie, the ring announcer, he was in it. I did 122 recognisable people at ringside, celebrities, sports writers, all kinds of people – and the gangsters. I had Carbo in there, Frank Costello . . . the Garden people wanted to cut them out of the painting. The Mayor of New York was in there. I fixed it up. Those people looked at my painting and said, "Hmm, you know, the mayor in this looks an awful lot like Frankie Carbo." One flick of the brush . . . and Frankie had become the mayor. Ha!'

Not all dignitaries were so concerned about their associations. Some of them craved to be immortalised by Neiman, to be accepted outside the dark offices and bars from where they pulled the strings on other people's lives.

'I used to come to the fights at the Garden with Angelo Dundee. He'd come up from Florida with his fighters and I'd tag along. It was logical I'd fall into step with what was happening at the Garden. John Condon was the PR guy. Nobody liked him much. But he liked me, gave me a press pass. And I'd sit there at the fights and draw all these guys. Frank Costello'd be there. Every time. Never missed. He wore his Borsalino hat, camel-hair coat, chain-smoking his cigarettes. A real, first-rate gangster. So I started making a sketch of him. And I swear he was posing for me. Didn't move at all.

'Jimmy Cannon was there. Big-time sports writer. He comes over to me. He says: "Are you crazy? You drawin' Costello? Does he know it?" "How do I know?" "Don't draw Costello. Costello wants no photographs!" "This is no photograph." "You're crazy!"

'Anyway, after the next fight, Teddy Brenner [the Garden's matchmaker] comes over and whispers to me, "Costello wants the drawings." So he didn't mind, apparently.

'Next week, Brenner comes up again and says, "Where's the drawing? You didn't sell it, did ya?" "I'd never sell it." "Well, Frank wants it." "I don't want to sell it to Frank Costello." 'And he says, "You crazy?"

'A couple of fights later, a guy, Margolis, who had a restaurant near the Garden, a rough place, comes up. So, Teddy Brenner, Harry Markson [the Garden's main promoter], John Condon, they say, "We'd like to invite you to dinner tonight at the restaurant with Mr Costello. He wants that drawing."

'So we go to dinner with Costello. He's got his wife. I got a girl with me. The conversation goes on for a while but nothing important. And then Teddy says, "How about the drawing, LeRoy?" Frank Costello couldn't hardly speak. He was dying of cancer of the throat. So he leaned over and said in this real husky voice, "I'd like to see that drawing. I'd like to have it." "Well, I haven't got it."

'The evening ended up with him not asking again. But there we are, sitting at the table with this big-time gangster. Everyone was watching the table, real nervous. Anyway, he died about a month after that. He never did get the drawing. He was a great guy. He was a celebrity.'

LeRoy was excited about this episode. As we chatted, he told me he was expecting a guest, 'a real-life gangster', who'd done two stretches for various crimes, eight years and four years, and he ran strip clubs now. 'He's a colourful guy. Likeable. A nice guy. I love him.'

I didn't doubt it. This was part of the fascination with crime and criminals I had stumbled upon. It was like New York of a century and more before, when Harry Thaw shot Stanford White in the Garden. Gotham was scandalised. And titillated. They love their villains.

Not many people in this story had a bad word to say about these people, men who killed other men, men who ruined the lives of other men (rarely women), men who, quite plainly, weren't going to heaven. Maybe it was part of the overall adoration of fame. Doesn't matter what you're famous for in New York. As long as you can pick up the tab or get someone to pick it up for you. Clout is measured in how long someone will wait for you, how much you can push it.

'I remember going up to Harry's office. I was going to do a life-size painting of Sugar Ray Robinson. Harry's office, on the second floor, looked down on Eighth Avenue. Ray Robinson was late. We expected that. Then up comes this pink Cadillac convertible. Out gets Sugar Ray. White jacket. He had his manicurist with him, and his barber, all these people. He didn't need 'em every day! But he liked to travel with his entourage. People just converged on him.

'He came up and I drew him. Things were so different then. No heavy security. He just walked right in. Before the fights, you'd go in the dressing room, no problem, see the fighters, talk to them and all that stuff.'

LeRoy remembers Toots Shor's, the bar near the old Garden that is gone now but gave birth to many tales, most of them authentic, the rest self-serving. 'Toots had his aisle seat at the Garden. Ed Sullivan sat nearby and Joe Louis would sit wherever he wanted. You'd just give up your seat for Joe, even if you'd paid a lotta money for it. The big thrill was when Joe Louis would walk over and say, "Howya doin', LeRoy?" And people'd look over with admiration.

'The thing about New York at that time was all the fight guys lived here too. If they weren't born here, they'd come in. Like Jack Dempsey and Gene Tunney. It'd have to be a big fight but sometimes you'd have Dempsey on one side of the ring and Tunney on the other. Front-row seats, of course. And every night they'd bring them in the ring and introduce 'em. John Condon would line the old champions up beforehand, he'd give a signal to the announcer Johnny Addie and they'd all file in like a chain gang. They were never forgotten, these guys. There was one guy . . . they wheeled him up the ramp in a wheelchair . . . what was his name? I don't remember. But they were all great guys.'

On 10 June 2007, LeRoy stepped up into the ring reserved for the game's greats. His health was on the slide, but he sent a message to be read out at the induction ceremony at the International Boxing Hall of Fame in Canastota, NY.

The IBHOF president Don Ackerman did the honours for him in his absence: 'To be included among these great fight figures and to be accepted into their circle is very meaningful to me. I've tried to measure up to the class, originality and style of the greats in my work . . . How lucky I am to be acknowledged along with proven heroes of boxing and to receive this recognition by the Boxing Hall of Fame. I am proud to be included as an honoree in this noble art.'

There wasn't a pug in the place who would begrudge the old poseur his place alongside them.

Bert Sugar was there soaking up the sun that day. He'd already made it onto the famed list. In what other line of legitimate baloney could a couple of wonderful old geysers like LeRoy and Bert move unchallenged alongside genuine fighting legends such as Roberto Duran, Carmen Basilio, Carlos Ortiz, Masahiko 'Fighting' Harada, José Torres, Emile Griffith, Gene Fullmer,

Ruben Olivares, Ken Norton, Alexis Arguello, Matthew Saad Muhammad, Marvelous Marvin Hagler, Ken Buchanan and Pipino Cuevas?

Bert, like LeRoy – actually, like a thousand guys in the fight game – invented himself to very good effect. That's not to say he doesn't know what he's talking about. He is one of boxing's more reliable historians. And, even more than Neiman, he has the ability to laugh at the absurdity of his own persona.

We meet in a bar at the Garden – not what it meant even forty years ago. New Garden. New bar. But Bert's stories are minted. He came from out of town but stayed and fell in love with New York, like millions before him.

'First of all the town of the fifties and sixties was different, you have to know. I mean, you could walk along Broadway at any hour of the morning and everything was on, everything was open, it was an all-night thing. It looked like the opening scene of *Guys and Dolls* and everybody walked the city and it just was a fun town. And it was still a big, big newspaper town. I mean, in the fifties I think there were still eight papers and there were so many editions. People would go out and get them, particularly for the race results or the early-morning handicapping, and go to the kiosk and then go back to Lindy's or Dempsey's or wherever. And there were always tables, particularly at Lindy's: there was the comedians' table, the dancers' table, the writers' table, the boxing table. I don't think there were many insurance men but, whatever fraternity, they had their own area.'

How did Madison Square Garden fit into the cultural and physical landscape of New York City in the fifties?

'Well, it was all about the boxing. The *Ring* office was on the second floor in the mezzanine. It was a square building built for boxing. Hockey came later, there was no basketball out here, so it was built for boxing. And there was an entryway, a kiosk there where you got some crap they passed off as orange drink for five cents, and there was a ten-cent hot-dog stand. That was equally terrible. And there was an arcade and you walked in and there was a bank of phones before you even got to the gate. The bank of phones was jammed up on Friday nights with people making last-minute bets. All the bookies had the fucking phones. And you got the impression people went to the Garden on Friday night, every Friday night, without sometimes knowing until they got up there who the hell was fighting. It's like I'm going to the poker game. I don't know who's going to show up tonight but I'm going. So, it's Friday night, I'm going to the Garden.

'You'd get inside and it would be crazy. There was an announcer by the name of Harry Balogh and he wore a tux and he sort of looked like a little powdered penguin. And he had this wonderful way of making grand statements. Well, one night he goes, "Ladies and gentlemen, our wonderful pianist Miss Gladys will now play the Star-Spangled Banner for you." So somebody way up in the upper fucking deck calls, "Gladys sucks!" And Harry replies, "Nevertheless, Miss Gladys will still play the Star-Spangled Banner for you."'

There are so many variants of that story. But it's still a good one. And if it wasn't the Garden and Harry and Gladys, maybe it should have been.

'Another night, Harry did it again. It was during the war and we used to have regular collections at public functions for orphans, sailors, whatever. The whole place was very patriotic, as you can imagine, and they used to collect the coins in these cans they'd pass through the crowd. So, anyway, one night Harry announces to the crowd before a big fight, "Ladies and gentlemennnn! Before the main event, little ladies will be going around shaking their cans. Give till it hurts." Lovely. And it was all good fun. You could hear people screaming and hollering.'

Once the fighting got under way, everyone was an expert.

'And they were loud experts. There weren't riots, they were just loud people. Everybody had an opinion and was not afraid to voice it. But there was hardly ever any trouble. Not in the old, old Garden, except maybe at the end. And the night La Motta loses the decision to Robinson – the chairs were flying that night.' That was in 1945, long before Bert or his contemporaries got to the Garden, but the stories endure nonetheless.

And minor skulduggery was the norm.

'Nobody checked anything,' Bert says. 'Like half the time the ticket-takers were stealing the gate. You have a drink, then you just go and give them five dollars, you'd sit in a ten-dollar seat. You would know what gate to go to, and who to give how much to. There was an atmosphere, no question. And you didn't get many women there. Not many blacks, they didn't have the money, and Latinos, there weren't many of them. Joe Louis – his supporters up in Harlem listened to him on the radio. Hardly ever saw him throw a punch.

'I remember my first visit to the Garden. I was maybe twenty years old and I just stood there looking at everyone. I'd come up from Washington with my mates, who'd come to town to get laid and maybe get some cheesecake, and I left them under the clock at the Roosevelt. I said, "See ya later, guys, I'm going to the fights." And I came back later after the fights and they were still standing there under the fucking clock, not got laid. But they'd had their cheesecake.'

Sugar, who had a law degree, was writing copy for an advertising agency and earning spectacular money for the time. '$65,000 a year for writing the words for the Nestlé ads!' He tired of the job, thirsting for the excitement of the fight game.

'I punched the boss in the mouth and they threw me out. I think that's fair. This man had carried me longer than my mother, he put up with my shit.'

Life, he reckons, was more innocent then – yet still as bad underneath. 'The wise guys may have had the same sinister thoughts as they do now, but you didn't think they did. I mean it was Eisenhower, everybody built bomb shelters to stop nuclear bombs. A lot of that changed – maybe some of our innocence was lost with the assassination of Kennedy. We no longer believed what we were told.'

Sugar did some journalism back then that took him to the edges of the Mob world. He remembers how the Mafia sweated through various judicial investigations and how the heat was really turned up on the Garden. 'Not a lot changed until they broke up the Garden. They put Carbo away for interstate trafficking and Blinky went.'

That left Jim Norris and Truman Gibson. Sugar confirms the view of Norris as the well-liked, well-turned-out hood. 'He had the connections, everywhere, and the venues, not just New York. In fact, he owned three of the six National Hockey League franchises. He owned Toronto, he owned Detroit and he owned Chicago. He buys the Garden and gets the television contract from NBC and he loves to hang, as many did, with notorious people, whether they were Mob or a Mickey Spillane-type of bad guy. He had the money but couldn't make the matches. That was up to Carbo and the boys. There was an octopus effect, and it became so because of Norris.'

But the man who instilled fear, who made sure things happened, was Carbo, the Prime Minister.

'Carbo, Mr Gray. All he'd have to do is look at you with those steel-grey eyes. Most of the writers, if he looked over, we'd just go, "The eyes." He was very, very handsome, always wore a nice suit, very slim, and walked with the authority of a man who has whacked a couple of people down and knows that, if he had to, a couple more would be whacked. The funny thing is, I once shook hands with him way back. I went up and said hello and he was nice, said hello, that was it. I mean, his guns are over there and they sort of looked at me. And when I got back to my seat somebody said, "You're a made man, are you?" I went, "No, I'm a kid. What the fuck is this made man shit?" I only found out later what it meant, that you were part of the Mob.

'If you went to a restaurant in this town in the fifties – instead of smoking and no smoking, in the Italian restaurant you had the no-shooting section. Most of the hits were in the fucking Italian restaurants.'

What about Toots Shor's then: was it as wild as they say?

'They tell a story – and I was not privy to this – Toots is talking to Dr Fleming, who invented penicillin – and Toots has no fuckin' idea who he's talkin' to. One of his waiters comes over and says, "Excuse me, Toots, Bill Rigney [the manager of the New York Giants at the time] just came in." So Toots turns to this stranger he's been talking to and says, "Excuse me, sir, someone important just came in." Ha! So he's talkin' to a fuckin' baseball manager rather than the inventor of penicillin.

'Then there's the time Charlie Chaplin's standing in line waiting with Paulette Goddard for a table. So he complains to Toots and Toots says, "I'm busy Charlie, get back in line." I used to go there all the time. That's how I became a sports writer. I was enamoured. I'm not a New Yorker, so I come here [from Washington] to be in advertising, but I really wanted to be a sports writer, just like [Howard] Cosell. So, watching these guys, I decided to wear a hat. Back in the fifties, they all wore hats. All of 'em. So I said to 'em, "Why you wearin' a hat?"'

They told him. It was, Bert recounts, to stop filings from the hot-metal production room slipping through the floorboards and landing on their heads in the newsroom below. Sugar believes that. Which is fair enough. He's good enough to acknowledge that sticking 'Press' in the hatband, as seen in the movies, was an affectation. The drinking, though, was for real.

Especially at Toots Shor's. Sugar says the reason the place closed was Toots was too generous with the free booze for famous clients. He'd put tables aside for Joe DiMaggio and never charge Joe or his friends for their drinks. They were magnets.

'When I started going there,' Bert recalled, 'drinks were two dollars and the bartenders had a cute habit of taking a dollar, putting it in their pocket and putting a dollar in the cash register. And one time I was seated about six seats away and I was watching a bartender put both dollars in his pocket and Toots went, "What's the matter, we're not partners any more?"'

Jackie Gleason, whose fame as New York bus driver Ralph Kramden in hit fifties TV series *The Honeymooners* outstripped all of his other comic work, was a regular. And regularly rat-arsed.

'Toots and Jackie Gleason, they're drinking all morning,' remembers Sugar, 'and they bet a hundred bucks on a race around the block. And Toots was fatter than Jackie. These are Toby mugs. So now Toots has got to go that way out the front door, once around the block, and Jackie's gonna go the other way. Gleason comes panting back after fifteen minutes and there's Toots sitting at the bar drinking, and Gleason pays him. After a couple more drinks, though, something occurs to him: he's running that way, and I'm running that way – why the hell didn't I pass him? Well, Toots had taken a cab . . . These things went on all the time.'

Sugar got hooked on Toots Shor's. 'Bernard Shor, his name was. Had been a bouncer in speakeasies. He came from Philadelphia and he was underwritten, some say by the Mob and I wouldn't be surprised, cos they ran all the speakeasies. All the stars had a spot in there. DiMaggio, Sinatra. Sometimes Gleason would be flat out, face down on the floor, and Toots would just step over him. The fight writers would all get in there too. And, yes, they'd have their envelopes from the Garden, their pay-offs. I used to call the old writers Sir Lunchalots.They'd take home doggy bags, through snowstorms. They hadn't eaten in months some of those guys. They knew what was going on. They knew the fixed fights.'

Ike Williams, briefly a world champion, had a right hand so

concussive he is regarded as among the hardest punchers in the history of boxing. He fought, and some times beat, the leading lightweights of his era – Beau Jack, Johnny Bratton, Kid Gavilan, Freddie Dawson, Jimmy Carter and, infamously, Chuck Davey. It was Ike's serious misfortune to be managed by Blinky Palermo, who robbed him blind – and, according to the fighter's testimony, persuaded him to take a dive against Davey in 1952. Not everyone believed Ike, however tempting it was to pour scorn on Palermo.

'Ike Williams? He didn't throw a fight. How could he? They hardly paid him any money in the first place. Blinky Palermo would just take his money, and that was it. Blinky did all his business from a deli in Philadelphia, Samson Street between Chestnut and Walnut Streets, believe it or not.'

The marquee fighter the Mob had to do business with after Louis retired was the man who knocked him out, Rocky Marciano. He suffered from being Italian, like most fighters of that heritage, but there can be no doubt the guys from the IBC were close to him. They would not have dared suggest Marciano sell his title or throw a fight but they spent a lot of time in his shadow. Sugar has a take on Marciano.

'One time Carbo, who, theoretically at least, had a piece of Rocky, offered him a thousand dollars for showing up at a banquet and then gave him a cheque. Rocky, who was tight with a capital T, said, "I don't want a cheque, give me five hundred dollars in cash in a paper bag." Carbo duly did – and Rocky left it in a cab.'

And that is a true story.

19

THE RECKONING

You know, if things pan out, these characters might gather one sunny day at Jacobs Beach, wander over to the bar at the Forrest Hotel, have a drink or two, light up a few cigars, maybe, and reminisce. Like old fight guys used to do.

Through the blue fug that drifts through the bar you might see them, battered or beautiful, rich or poor, talking about how lucky they were to be part of the craziest business there ever was. Or not so lucky. They all have a story. They all came through it in their own way, some better than others . . .

There's Frankie Carbo. How lucky was he, in the end? When he went down for twenty-five years in 1961, Judge George Boldt told the court, 'Never in his life has he been associated with any useful activity. He has been a menace to humanity and a hardened and degenerate criminal.' He was released on compassionate grounds in 1976 because of illness, and died on 9 November that year, aged seventy-two, of heart disease. Carbo left nobody behind who missed him for more than a minute.

Blinky Palermo, as ever, is standing beside his pal. He did seven and a half years for the Jordan–Akins scam and faded from view. He died where he lived, Philadelphia. Not only did not many know of his death, in 2006, aged ninety-one, few people knew he was even alive, the original man in the shadows.

There in the corner, that's Frank Costello. On 2 May 1957, Vincent 'The Chin' Gigante went to the lobby of the Majestic, the fancy apartment block favoured by the Mob, overlooking Central Park, and shot him in the head (not long after the 'Prime Minister' got out of prison) on behalf of Vito Genovese. Frank didn't die, though – and, after making up with Vito, would carry on his underworld business quietly from his Manhattan eyrie until he died of a heart attack on 18 February 1973.

A heart attack took down Jim Norris, too, in the city in which he was born, Chicago, on 25 March 1966. He was fifty-nine. After the US Supreme Court broke his hold on the IBC in 1959, he went back to the racetrack, looked after his hockey team in Chicago, as well as stadiums there and in St Louis, and ran companies that dealt lucratively in grain, real estate, cattle and railroads. Norris, like Truman Gibson, clung to a certain invalid integrity.

Truman, looking every inch the lawyer, is in a huddle with Joseph McCarthy, quizzing the infamous senator and red-baiter with lawyer-like intensity. Gibson got out of boxing at the end of the fifties and went back to practising law in south Chicago almost until the day he died, quietly on 23 December 2007, two years after his wife's death. He was ninety-three. He put out a ghosted autobiography, in which he talked about all of the good and little of the bad in his life. A decent, intelligent man, Truman, briefly, went wrong.

Joe McCarthy's hounding of the left ruined the working lives of scores of writers, directors and actors, some of whom fled to the UK, where they wrote episodes of the TV show *Robin Hood* under assumed names. The Senate castigated McCarthy for his 'reprehensible' behaviour. He died of hepatitis, exacerbated by his excessive drinking, on 2 May 1957 – the day Frank Costello got shot.

That's Mike Jacobs behind him, drinking water. When the IBC eased him out of the Garden in 1949, Uncle Mike and the

Twentieth Century Sporting Club were history. He retired to his Jersey mansion, returning very occasionally to his old workplace; he was often greeted warmly, although he found it beyond him to crack even a smile in response. 'It can't be a broken heart,' someone said of his sour demeanour, 'he hasn't got one.' Jacobs died in January 1953. His 'beach' lived on a few years, until it disappeared even as a notion, sometime in the mid-fifties.

Jimmy Johnston didn't much care for Jacobs. The witty, tricky little Scouser had run boxing's HQ with panache between 1931 and 1937, before Jacobs, with the help of his pals in the underworld and newspapers, saw the 'Boy Bandit' off the premises. Johnston had arrived in New York in the roaring nineties and liked people to know he'd been a pugilist. In fact, Jimmy had just one fight, against another debutant called Patsy Donoghue, at the Lenox Athletic Club in New York on 14 March 1899. He was knocked out in the tenth – and retired. Thereafter, he managed Mike McTigue, Ted 'Kid' Lewis and, after the Garden gig was up, Archie Moore. In May 1946, Jimmy made a comeback as a promoter. While watching a light-heavyweight in whom he had an interest get battered at St Nick's, Jimmy, who never smoked or drank in his life, said of the performance, 'It made me sick.' He turned pale, collapsed right there at ringside and died of a heart attack. He'd mixed with everyone in boxing from John L. Sullivan to Joe Louis. He was seventy.

Damon Runyon sits, cogitating, sipping on his coffee, eyes darting into the corners for stories he could relate from this little gathering. He knows them all. This is his place, and town, the New York he helped invent. The writer who was born in Manhattan (Kansas) died, aged sixty-six, in Manhattan (the big one) and had his ashes scattered over the city by the famous flyer, Eddie Rickenbacker, ten days after throat cancer got him on 10 December 1946 (just seven months after Jimmy Johnston). That week in the Garden he loved, the outstanding Jersey welterweight

Charlie Fusari was knocking out Chuck Taylor in seven in the main event, and Kid Gavilan beat Johnny Williams over ten. The usual faces were ringside, the Jersey mobsters, Runyon's old newspaper pals. And they said some nice things about him when they repaired to Toots Shor's.

Toots Shor left behind not a bar but a vivid memory of one, full of great stories. Shor's eponymous establishment was very much of its time. It is inconceivable to imagine a bar owner in New York today who could so easily draw into his ambit the likes of John Wayne, Frank Sinatra, Mickey Mantle and Marilyn Monroe alongside the underpaid hacks who chronicled their deeds. And, once that time had passed, the proprietor was lost without his stage. The lustre started to fade in many ways, from bar takings to the clientele. Joe DiMaggio, one of his best friends for many years, cut Toots from his circle when he made a disparaging remark about Monroe. It was a rare social defeat for Toots. In the end it began to parody itself, a photo opportunity for tourists. He sold up in 1971 and reopened around the corner at 33 West 52nd Street, but it was never the same. Toots had become mired in tax difficulties, owing $269,516 when he closed again. One final go at recreating the old days, at 5 East 54th Street, failed – and that was it. Toots died on 23 January 1977, aged seventy-three.

And what about the other Jacobs, Joe, New York to the tips of his spats, born, bred and buried? 'Yussel the Muscle' is the author of three memorable lines: Two-Ton Tony Galento's famous threat to Joe Louis, 'I'll moider da bum!' and his excuse on behalf of Max Schmeling, 'We wuz robbed!' after he lost to Jack Sharkey in their second world title fight. Jacobs also said, 'I shoulda stood in bed,' but, whatever the folklore, it was not at a fight but a baseball match in Detroit in 1935. Jacobs died of a heart attack in May 1940, aged forty-three, and was widely mourned.

Estes Kefauver, as ever, looks a little out of place. He keeps his distance, like a headmaster coming across boys smoking behind the shed. After busting the IBC, the Keef carried on campaigning against the evils of the world until, one day in August 1963, he staggered from the floor of the Senate, complaining of an upset stomach. He died the next day of a ruptured heart artery in Washington's Bethesda Naval Hospital, aged sixty.

Joe Louis stands quietly, centre ring, you might say, where he ought to be. Once the charade of his working for the IBC was out of the way, he took to golf and a bit of wrestling (in the ring and with his parlous finances). He gave an interview in 1972 in which he said, 'Since 1969, I've had a little trouble with the Mafia. At one time my life was in real danger. They tried to put me out of the way. I wasn't sick at all when they put me in that hospital in Detroit . . . er, Denver. The Mafia put pressure on my wife to put me in the hospital. But it'll be straightened out. I'm not worried at all any more.' Asked if he felt safe, he added, 'Oh no. They might come back at any time. One thing I know. Your best friends can set you up. Your best friends can kill you.' Joe died of a heart attack in Desert Springs Hospital on 12 April 1981. Nine days later, he was buried at Arlington National Cemetery with full military honours, recognition of the service he gave to his country, and a second-rate apology for the way the government hounded him over taxes and into bankruptcy and mental distress.

James J. Braddock stands beside Joe, in his shadow, as ever. He retired in 1938 after beating Tommy Farr, enlisted in the US Army in 1942 and saw action in the South Pacific. He had a construction company after the war, and lived happily with his family in North Bergen, New Jersey. He died peacefully in 1974, aged sixty-nine. In 2005, his life and career were immortalised, with embellishments, in the movie *Cinderella Man*.

And there's Max Schmeling. He struggled after the war, but did pretty well for himself when offered a Coca-Cola franchise in Germany. He met Louis again in the fifties and sealed their friendship by helping him financially. Schmeling is remembered in books and a movie, but shunned publicity. He lived quietly in Hamburg with his wife of fifty-four years until 31 January 2005, when he failed to come out of a coma brought on by a heavy cold. He was ninety-nine.

Coley Wallace is part of this heavyweight story, if on the margins. Which is where he stands now, elegantly dressed but not engaging his fellow fighters, more comfortable in the background. He had much in common with Joe, Jimmy and Max. He was a boxer too, but he missed out on a big part in the story, partly because he wasn't quite good enough, partly because he was the sort of guy who didn't push himself. He'd been good in his younger days – good enough to beat Rocky Marciano in the Golden Gloves at the Ridgeway Arena in 1948. It was the last fight Rocky ever lost. But Coley was reduced to a walk-on part in the bigger scheme of things. It was his dreaded luck to be managed by Blinky Palermo, and he was seen not as a contender but as a pawn in the bigger game. He claimed Palermo not only paid him just $3,000 of a $20,000 purse for fighting Ezzard Charles in San Francisco in December 1953, when he was stopped in the tenth, but that he was drugged beforehand. 'I get angry when I think about Blinky Palermo,' Coley said once. 'He could have done better by me. He ruined boxing for me.' After he quit boxing, he ended up playing the Brown Bomber in the fifties Hollywood movie, *The Joe Louis Story*. It never led to a high-profile career in Hollywood, although Martin Scorsese cast Wallace as Louis in *Raging Bull*. He liked to reminisce, quietly, with fellow ex-fighters at Ring 8 in New York, where everyone remembered him as a gentle man, and a gentleman. Coley died in New York, in modest circumstances, the day before Schmeling died in Hamburg. He was seventy-seven.

Tommy Farr, the Tonypandy Terror, is threatening to sing. He stayed on in America after losing to Louis in '37 and had four fights in the Garden, losing on points to Jimmy Braddock, Max Baer, Lou Nova and Red Burman. He came home, disillusioned by continued approaches from the Mob and gamblers. He boxed on, moved to the south coast, opened a restaurant, which failed, and extended his career to pay off debts before retiring in 1953, without ever reaching the heights again. He had a fine singing voice, recorded one or two songs, but he struggled with his finances. He wrote a column in the old *Sunday Pictorial* until 1961. The best heavyweight Wales ever produced died on 1 March 1986 – St David's Day – aged seventy-one.

Throughout his career, Sugar Ray Robinson resisted the Mob. He was, uniquely perhaps, untouchable. After his fights with La Motta, he lost and won against Randy Turpin and fought on, as is the fighter's curse, too long. He grabbed at history by attempting a first and moving up to light-heavyweight to challenge Joey Maxim. Ray retired through heat exhaustion after thirteen rounds. Maxim remarked, 'What did I have – an air conditioner?' Robinson won and lost the middleweight title three more times and retired in 1965 at forty-four. In a life marked by incredible highs and a few lows, he danced onstage, romanced and entertained, briefly was a burning star on TV, lost his money and suffered mentally. He died on 12 April 1989, aged sixty-eight.

Rocky Marciano, as everyone knows, died in a plane crash, in 1969, the night before his forty-sixth birthday. He left behind the cleanest record of any heavyweight (the only champion in the history of the division to retire undefeated), and a stack of money that remains undiscovered to this day. Rocky was notoriously tight. He quit boxing not because he'd had enough, but because he wanted to spite his manager, Al Weill. The Rock begrudged anyone making money out of him. Archie Moore was one of only two fighters – alongside Jersey Joe Walcott – to knock him down – so Rocky gets the Ol' Mongoose to order the drinks. He takes an occasional glance at Carbo and the boys. They look back, smiling knowingly.

Rocky Graziano is as ebullient as ever, unapologetic and

mischievous. When he left boxing, it was with his reputation as one of the sport's most murderous punchers intact. 'No one ever hit me harder,' Sugar Ray Robinson said. Always a joker, he forged a minor career as a TV comedian with Henry Youngman, and was a regular face on other TV shows of the fifties. He took to painting too, and wrote – or had written for him – an autobiography, *Somebody Up There Likes Me*, which was turned into a film in 1956, with Paul Newman, which pleased him no end. Rocky had a couple of minor acting roles, notably as an ex-pug in Frank Sinatra's 1967 film, *Tony Rome*. He died on 22 May 1990, aged seventy-one.

Rocky nods to Cus D'Amato, who managed him once but wasn't with him when he won the title; Cus, grumpy as ever, ignores him. D'Amato, who slept with a gun, so paranoid was he about the Mob, knew more than he ever let on about the gangsters, and there were some who wondered if he really was that scared of them, if he didn't actually have connections. Cus, of course, finally got his world champions, Floyd Patterson, José Torres and Mike Tyson – although he died, on 4 November 1985, before Tyson began his reign of terror. Cus was seventy-three.

Billy Graham, Graziano's New York contemporary and cut from different cloth, lost, but won. So seared in the collective consciousness was the unfairness of his defeat in the Garden by Kid Gavilan in 1951 that Billy will forever be remembered as the Uncrowned Champion. He lost to the Kid again and retired in 1955. In all Billy's 126 fights, Gavilan was the only opponent to put him down. Once he quit the ring, Graham worked as a referee and as a representative for a liquor company. He died of cancer at his home in West Islip, Long Island, on 22 January 1992, aged seventy.

Kid Gavilan, typically, looks slick, so cool. He is the buzzing centre as he regales a little knot of fighters around him. But the Kid did not always tell them the whole story. A couple of years before

Gavilan died, of a heart attack in Miami on 13 February 2003, aged seventy-seven, a British journalist, John Duncan, met Gavilian's second wife, Olga, in Havana. Gavilan tried to bring her to America in the sixties but a maze of red tape strangled the venture. She seemed to bear him no ill. And she corrected the long-held myth that he'd run out on the revolution. 'He gave them money,' Olga said. 'He wanted to retire here.' But they stripped him of what he had and he left – unable to bring Olga with him. He contacted her again in 1985 and said he could get her a visa. 'I told him I couldn't leave,' she informed Duncan. 'It was so difficult. I had made a new life for myself. In 1985, I asked him for a divorce. We'd been married seventeen years and there was no future. Cuba got in the way.'

Don Jordan, never an angel, mixed with mobster Mickey Cohen and others. It caught up with him in the end. Jordan was robbed and badly beaten up in a parking lot in Los Angeles in September 1996. He went into a coma and died in a nursing home in San Pedro, California, on 13 February 1997, aged sixty-three.

Jackie Leonard, who survived the trauma of ratting on the Mob and having Carbo and Palermo put away, keeps his distance from the wise guys in the Forrest bar. After those eventful months in 1958, he returned to his first qualification, engineering, and worked on construction jobs in Saudi Arabia and Vietnam. Leonard died of heart failure in a nursing home in Winter Haven, Florida, in 2007, aged eighty-nine.

Joe Miceli was the original Mr Friday Night, so often did he appear at the Garden on TV in the fifties. He admits he was 'washed up' at the end, but he had no regrets. He lived quietly in Long Island, meeting up regularly with old boxing pals in Ring 8. He died in July 2008, aged seventy-nine. 'When we left him at the Southside Hospital [in Bay Shore],' his son Joe Jr said, 'he looked and felt the best we'd seen him in months.'

Not all of our players have exited the stage . . .

When Virgil 'Honey Bear' Akins lost to Jordan in December '58, he did not have even a good pay day to soften the blow. Immediately after the bout, an official from the Internal Revenue Service informed him the government was taking $22,599 from his purse of $28,423 in back taxes. It was clear Akins was trying in the rematch, as Palermo sat ringside in an agitated state throughout. Jordan got the decision again, and Virgil's career fell away after that. He lost eleven of his last twenty-two fights. Nothing has been heard of Akins since he retired from boxing in 1962.

Budd Schulberg is still kicking – at least he was the last time I saw him, at the Joe Calzaghe–Bernard Hopkins fight in Las Vegas in 2008. He was ninety-four and frail, but still sharp as a Willie Pep jab.

Jake La Motta? At the time of writing, he's still with us, up in his Manhattan apartment, sitting by the phone, waiting for the next offer from a writer to talk about his seven wives, Sugar Ray, the Mob, whatever. They haven't put him down. Not yet.

GOOD NIGHT, AND GOOD LUCK

The above line was the signature farewell used by the journalist Edward Murrow, who did so much with his television colleagues at the CBS in New York to expose the insidious evil of McCarthyism in the fifties when American law-makers and enforcers would have been better employed hunting down gangsters rather than a few old leftist writers. And I just couldn't think of a better way to sign off. Murrow 'threw stones at giants', as one fan said, and we should all be grateful to him. Without Murrow and Estes Kefauver, a lot of bad things would have gone unpunished in the fifties.

Muhammad Ali, with Richard Durham, *The Greatest, My Own Story* (Book Club Associates, 1976).

Dave Anderson, *In the Corner – Great Boxing Trainers Talk About Their Art* (Morrow, 1991).

Teddy Atlas and Peter Alson, *Atlas, From the Streets to the Ring: A Son's Struggle to Become a Man* (HarperCollins, 2006).

Richard Bak, *Joe Louis, The Great Black Hope* (Da Capo, 1998).

Jim Brady, *Boxing Confidential – Power, Corruption and the Richest Prize in Sport* (Milo Books, 2002).

Teddy Brenner, as told to Barney Nagler, *Only the Ring Was Square* (Prentice-Hall, 1981).

Jimmy Breslin, *Damon Runyon, A Life* (Hodder & Stoughton, 1991).

John Dickie, *Cosa Nostra, A History of the Sicilian Mafia* (Hodder & Stoughton, 2004).

Gavin Evans, *Kings of the Ring – The History of Heavyweight Boxing* (Weidenfeld & Nicolson, 2006).

Nat Fleischer, ed., *The 1960 Ring Record Book and Boxing Encyclopedia* (O'Brien Suburban Press, 1960).

Charles L. Fontenay, *Estes Kefauver, A Biography* (University of Tennessee Press, 1980).

Truman K. Gibson Jr, with Steve Huntley, *Knocking Down Barriers – My Fight For Black America* (Northwestern University Press, 2005).

David Halberstam, *The Fifties* (Fawcett Books, 1993).

Peter Heller, *In This Corner – Forty World Champions Tell Their Stories* (Robson Books, 1985).

Brian Hughes and Damian Hughes, *Peerless, The Sugar Ray Robinson Story* (Cromwell Press, 2007).

Sam Kashner and Jennifer MacNair, *The Bad and the Beautiful – A Chronicle of Hollywood in the Fifties* (Time Warner Paperback, 2002).

Estes Kefauver, *Crime in America* (Doubleday, 1951).

Jake La Motta, *Raging Bull – My Story* (Da Capo, 1997).

George J. Lankevich, *New York City, A Short History* (New York University Press, 1998).

A. J. Liebling, *The Sweet Science* (Simon & Schuster, 1951).

A. J. Liebling, *A Neutral Corner, Boxing Essays*, Fred Warner and James Barbour, eds (Northpoint Press, 1990).

Just Enough Liebling, Classic Work by the Legendary New Yorker writer, introduction by David Remnick (North Point Press, 2004).

Joe Louis, with Chester L. Washington and Haskell Cohen, *My Life Story* (Eldon Press, 1947).

David Margolick, *Beyond Glory – Joe Louis vs Max Schmeling, and a World on the Brink* (Alfred A. Knopf, 2005).
Mike Marqusee, *Redemption Song – Muhammad Ali and the Spirit of the Sixties* (Verso, 1999).
Jim McNeill, *That Night in the Garden – Great Fights and Great Moments from Madison Square Garden* (Robson Books, 2003).
Jürgen Müller, ed., *Movies of the 50s* (Taschen, 2005).
Michael Munn, *Sinatra: The Untold Story* (Robson Books, 2001).
Patrick Myler, *Ring of Hate – The Brown Bomber and Hitler's Hero, Joe Louis v Max Schmeling and the Bitter Propoganda War* (Mainstream Publishing, 2005).
Thomas Myler, *Boxing Hall of Shame – The Fight Game's Darkest Days* (Mainstream Publishing, 2006).
Barney Nagler, *James Norris and the Decline of Boxing* (Bobbs-Merrill, 1964).
Jack Newfield, *Only in America – The Life and Crimes of Don King* (Morrow, 1995).
Harry Otty, *Charley Burley and the Black Murderers' Row* (Tora Book Publishing, 2006).
Allen S. Rosenfeld, *Charley Burley – The Life and Hard Times of an Uncrowned Champion* (1st Books, 2000).
Jeffrey T. Sammons, *Beyond the Ring – The Role of Boxing in American Society* (University of Illinois Press, 1990).
Jeff Silverman, ed., *The Greatest Boxing Stories Ever Told – Thirty-six Incredible Tales from the Ring* (Lyons Press, 2002).
Everett M. Skehan, with family assistance by Peter, Louis and Mary Anne Marciano, *Rocky Marciano* (Robson Books, 2004).
Russell Sullivan, *Rocky Marciano, The Rock of His Times* (University of Illinois Press, 2002).
Anthony Summers, *Official and Confidential: The Secret Life of J. Edgar Hoover* (Pocket Books, 1993).
Mark Thornton, *Policy Analysis: Alcohol Prohibition Was a Failure* (Online, 1991).

Burton B. Turkus and Sid Feder, *Murder, Inc., – The Story of the Syndicate* (Da Capo Press, 1992).
Dan Wakefield, *New York in the Fifties* (Houghton Mifflin/Seymour Lawrence, 1992).
Geoffrey C. Ward, *Unforgiveable Blackness: The Rise and Fall of Jack Johnson* (Pimlico, 2004).
Stanley Weston, ed., *The Best of the Ring – The Bible of Boxing* (Bonus Books, 1996).

People invariably asked if writing a book about mobsters was in any way perilous, given the neighbourhoods it was necessary to frequent. And the answer was an emphatic no. It is surprising, and disturbing, how normal and nice some 'bad' people can be.

There are times in your life when you struggle to believe you're not dreaming – and nights with Chuck Wepner fall into that netherworld. When he calls himself a legend, there is no argument in New Jersey, except maybe in his own house in Bayonne, the archetypal Mob town. His latest wife, Linda, might have given Muhammad Ali an even tougher argument than Chuck did in 1975.

According to Wepner the fight, and the rest of his life, inspired Sylvester Stallone to make the *Rocky* movies – but this was not properly acknowledged until after a legal squabble over money and recognition.

Chuck became famous as the Bayonne Bleeder, a fighter with more leaks than Watergate, and strides through his manor like a king. But he always kept his lip zipped – which the Mob appreciated after he'd gone down on cocaine charges. Chuck was not only a generous host – who opened doors upon which only a vicar would knock with impunity – but he knew the stories. He shared some of them with myself and my adventurous friend, Mark Collings, and led us into the mysterious fringes of Mobworld.

Others spoke freely about their associations and recollections: Gil Clancy, Budd Schulberg, Angelo Dundee, Al Certo, Bert Sugar, Thomas Hauser, Howie Albert, Teddy Atlas, Lou Duva, LeRoy Neiman, Pete Hamill, Joe Miceli, José Torres, the guys and dolls in Irene's bar . . . and some we've no doubt forgotten.

The cast from the past is longer still: Damon Runyon, Bat Masterson, Doc Kearns, Tex Rickard, Stanford White, Sam Langford, Jack Dempsey, Jack Johnson, Owney Madden, Bugsy Siegel, Joe Louis, James J. Braddock, Joe Gould, Max Baer, Maxie Rosenbloom, Two-Ton Tony Galento, Al 'Bummy' Davis, Billy Graham, Kid Gavilan, Rocky Marciano, Rocky Graziano, Sandy Saddler, Willie Pep, Lulu Perez, Ike Williams, Jimmy Walker, Dan Parker, A. J. Liebling, Barney Nagler, Teddy Brenner, Jimmy Cannon, Mike Jacobs, Marshall Stillman, Cus D'Amato, Sonny Liston, Blinky Palermo, Elia Kazan, Arthur Miller, Walter Winchell, Frank Costello, Jim Norris, Truman Gibson (who, thankfully, I was able to meet before he died at the age of ninety-six) and Frankie Carbo . . .

There is no denying, either, the help of the Internet. I resisted the temptation, as far as is possible, to trawl the insidious beast but am indebted to one site known to nearly all boxing writers: boxrec.com. The site's creators deserve much credit for compiling such a comprehensive record, with time, dates and occasional embellishments of the deeds of nearly every fighter who laced up. The archives of the International Boxing Hall of Fame were similarly useful.

Newspapers of the day were a gold mine. For anyone who thinks the old dog is dead, that we are now married forever to the screen and keyboard, take an afternoon out to flip through some musty files, linger on the page and gaze for a while at the type and faded ink, the black-and-white photos, the cheesy headlines, the quaint advertisements – and then tell yourself how lucky you are to be transported back to a forgotten time. Among those I found helpful were my own newspaper, the *Observer*, and its sister organ, the *Guardian*, both of which have downloaded their wonderful archives, the *New York Times*, *Time* magazine, *Hollywood Confidential*, the *New York Post*, the *New York Daily News*, the *New York Herald Tribune*, the *Cleveland Plain Dealer*, the *Philadelphia Inquirer*, the *Chicago Tribune* and the *Miami Herald*.

I have trunkloads of ancient boxing magazines too that provided snippets and flavours of the day. *Boxing News*, the oldest and best, lit up a lot of dark corners. So did the *Ring* (although I developed a perhaps illogical suspicion of the magazine's long-time editor, Nat Fleischer, a man who, it seemed to me, was too close to his subject to stand in judgement so often and so loudly).

And then there was the editing, the process which Runyon and his pals considered a nuisance but, without which, I also might have fallen into patterns of ill-judged fantasy. For shaping the final product, then, I have to thank Tristan Jones, whose diligence might be rivalled only by a man trying to invent the perfect toothpick.

It has taken a while to get this far, for one reason and another, but I have enjoyed the journey. I hope you have too. If I've left anyone out, maybe you should have returned the calls. And if I've offended anyone with the capacity or inclination to simultaneously speak out of the side of his mouth and inflict gratuitous pain, I will be contactable only by carrier pigeon somewhere in the Azores.

Kevin Mitchell
North London, 2009

INDEX